AF353051

ION OTEȘCU

*

ROMANIAN
STAR & SKY LORE

English Translation, Foreword & Afterword by
Andrei Dorian Gheorghe & Alastair McBeath

Editura ASTROMIX
Cluj-Napoca
2024

Descrierea CIP a Bibliotecii Naţionale a României
OTESCU, ION
    Romanian star & sky lore / Ion Otescu ; English translation,
foreword & afterword by Andrei Dorian Gheorghe &
Alastair McBeath. - Cluj-Napoca: Astromix, 2024
    ISBN 978-606-95691-3-9

I. Gheorghe, Andrei Dorian (trad. ; pref. ; postf.)
II. McBeath, Alastair (trad. ; pref. ; postf.)

39

Original Romanian text written by Ion Otescu:
"Credinţele ţăranului roman despre cer şi stele"
Published in Bucharest, Romania, 1907.

This English translation was prepared by
Andrei Dorian Gheorghe and Alastair McBeath.

Andrei Dorian Gheorghe is Cultural Advisor (since 1995) and
Cosmopoetry Festival Director (since 1996) of the Romanian Society
for Meteors and Astronomy (SARM). He coordinates the website
Cosmopopetry - SARM and Friends or www.cosmopoetry.ro (since
2005) and has published over 1,000 astropoems and astro-photo-
poems in English on four continents, some of them in projects of the
International Meteor Organization, NASA, ESO, Science Fiction
Poetry Association and Astronomers Without Borders.

Alastair McBeath was for over two decades the Vice-President of the
International Meteor Organization during the 1990s and 2000s, and
ran the Meteor Observing Section for the UK's Society for Popular
Astronomy for nearly thirty years from 1983. He has also written
extensively on aspects of visual astronomy, mythology, folklore and
ancient history since the early 1980s. Although largely retired now, he
still maintains active interests in all these spheres.

All the interior illustrations were drawn by Alastair McBeath.
The cover was taken from Ion Otescu's original constellation map (1907).

Ion Otescu
(1859-1932)

Ion Otescu was born in the town of Buzău, Romania, on September 9, 1859, and died on March 10, 1932 in Bucharest.

He obtained his degree from the Mathematics Department of Bucharest University, graduating in 1886. After this, he continued living in Bucharest, writing books on mathematics, geometry and cosmology for school pupils, and making examinations of the calendar. During this time he was the director of a secondary school and a member of the Permanent National Council for Instruction. He later became the General Inspector of Schools (1912-1914).

Modernly, he is regarded as the founder of the study of Romanian astromythology through his publication "Credinţele ţăranului român despre cer şi stele" ("Romanian Star & Sky Lore").

CONTENTS

TRANSLATORS' FOREWORD

General notes

Ion Otescu's text is a remarkable document. It was first published as "Credinţele ţăranului român despre cer şi stele" (literally "Romanian Peasants' Beliefs in Stars & Sky") in 1907 in Romanian, the culmination of twelve years of research and investigation. In part, it preserves an oral tradition which, from the age of some of his correspondents, must descend to the late or mid 18th century at the very least, most likely a century or so before then, and probably still further. Whether it can be directly traced to the time of the Roman conquest of Dacia, and the subsequent Christianization of Dacia/Romania, as Otescu himself suggested, is far less certain. The importance of Emperor Trajan still in contemporary Romanian thought and legendary in the 21st century implies that tales of Trajan and events around the time of the Romano-Dacian wars in the early centuries C.E., were preserved orally for some considerable time, or were somehow revived at a time now sufficiently distant before the 17th or 18th centuries to seem to have thus survived. The re-use of many, but not all, the ancient Greco-Roman constellations by the Romanians, does imply a separate, more important, set of local traditions which kept such influences relatively at bay, or amended them while still retaining part of their nature, even though these were apparently not always the same in different places.

Otescu was working to compile his information during a pivotal time in Europe for such matters. In many places by the late 19th and early 20th centuries, there was a growing realization of just how much of such old traditions had been lost in the switch from predominantly rural, agricultural economies, to heavily urbanized, industrial ones, especially in northwest Europe, over the previous century and more. Elsewhere, this was a time when collections of all manner of folklore and local legends were being created by others who perceived what was happening too, and wished to preserve that information for posterity. Otescu was not alone in specialising in astronomical folklore, but looking through those other texts, perhaps the most comprehensive of which was R. H. Allen's "Star-Names and Their Meanings" of 1899 (reprinted by Dover Publications in 1963 in facsimile, as "Star Names: Their Lore and Meanings"), we see how very little information had survived even then of the old European constellation and star names, other than those from ancient Greek, Roman and medieval Arabic texts, or those subsequent attempts to fill gaps or amend the constellations for different purposes. In the British Isles, for example, the most comprehensive survey of surviving local constellation names, collected in the late 19th and early 20th centuries, is that by Marie Trevelyan ("Folk-Lore and Folk-Stories of Wales", Elliot Stock, 1909, primarily Chapter III). She was able to identify old Welsh names for a mere seven constellations, plus the Milky Way and the ecliptic! She did list a further 27 names for other objects in the sky, mostly probably constellations, but could not link them to specific star-patterns. Consequently, as far as we are aware, the material preserved by Ion Otescu in his work is unique in its completeness and depth among the surviving astromythologies of Europe, outside the Greco-Latin sphere.

To the best of our knowledge, this is the first English translation of Otescu's work ever published in print. We prepared an earlier version of the text in the mid-1990s, and hoped to publish it around the 90th anniversary of its original appearance. Unfortunately, we were unable to find a publisher then. We subsequently revised the text, and presented it as an online electronic version as part of the cultural activities during International Year of Astronomy 2009 (IYA; on the website of the Romanian Society for Meteors and Astronomy - SARM, linked from the official IYA 2009 website at National Nodes - Romania), just over a century after Otescu first published his book. Now, we are delighted to finally realise the translated work as a physically-printed book in its own right, and have taken the opportunity to completely check and revise the translation in doing so.

In our translation, we have chosen to concentrate on the Romanian notes and discussions, endeavouring to retain as much of the spirit of the work, and to remain as close to the original Romanian text in these aspects, as possible. We have retained specific Romanian terms wherever practical, but we have also given an English translation the first time each term appears in the work. In the original, there are some errors, and in places there are statements expressed curiously - occasionally archaically - and sometimes possibly offensively to modern eyes (the very low opinions of women in several places spring readily to mind), but we have usually translated these without comment.

Otescu often added discussions on elements of sky lore from various past cultures. We have abbreviated some of these, and omitted one or two, as the majority are available already elsewhere in English, and have done so especially where they added nothing to the Romanian material. He also provided footnotes, identifying which of his correspondents had given which name for a constellation or asterism, and adding occasional other remarks. Here, we have fitted details from the footnote remarks, where relevant, into the main text, and have added numbers indicating which of his correspondents provided which name, set inside squared parentheses, "[ ]". These numbers are the same as those given towards the end of the book in the "List of Sources Used". Where some clarification or correction was felt useful, including adding the official International Astronomical Union star designations for stars Otescu named (using either Greek letters for the brighter stars, or numerals for some of the fainter ones), we have inserted these into the text also in squared parentheses. We have kept our additions to a minimum, but would recommend that anyone interested in following-up on the non-Romanian myths should check beyond this text for best accuracy. A handy, referenced, guide to these is the revised, second edition of Ian Ridpath's "Star Tales"

(Lutterworth Press, 2018), for instance, while on the derivations of star names, Paul Kunitzsch and Tim Smart's "Short Guide to Modern Star Names and Their Derivations" (Otto Harrassowitz, 1986) is invaluable. We have also rearranged a few text sections to better fit the flow of the narrative.

α	alpha	ι	iota	ρ	rho
β	beta	κ	kappa	σ	sigma
γ	gamma	λ	lambda	τ	tau
δ	delta	μ	mu	υ	upsilon
ε	epsilon	ν	nu	φ	phi
ζ	zeta	ξ	xi	χ	chi
η	eta	ο	omicron	ψ	psi
θ	theta	π	pi	ω	omega

The Greek alphabet

One point of note concerns Otescu's use of the term "paper" (or occasionally "notice", or a similar word), which is not used in its modern academic sense as representing a published article in a journal, but refers instead to the papers Otescu received from his various correspondents outlining the myths and beliefs as known to them. Regrettably, we have not been able to establish what happened to these original documents after Otescu's book was published. If they still survive, they would be a valuable resource in their own right, especially as Otescu himself points to the possibility that other constellations than he discusses were mentioned in them, towards the end of Chapter One. This provides a tantalising glimpse of what has almost certainly now been lost.

Romanian diacritics and pronunciation

These are just a few brief notes on the pronunciation of Romanian words, for those unfamiliar with them. At a general level, most consonants should be crisply pronounced, and most vowels kept short. "A" is similar to the short "u"-sound in English "bud", "come", "put". "H" is never strongly aspirated, but "r" is always trilled, and "l" should be pronounced as it is in English when it precedes a vowel. "U" is always said as the "oo" in "soon", "moon". The various accented letters are pronounced as follows:

"ă" as the vowel-sound in "nurse", "hearse", "terse" in English;

"â" and "î" are both midway between the "ee"-sound in "keep" and the "oo"-sound in "moon" (we retain Otescu's original spellings here, although in modern Romanian, the "î" has been replaced almost entirely by "â" since 1993);

"ş" is pronounced "she"; and

"ţ" is pronounced "tse".

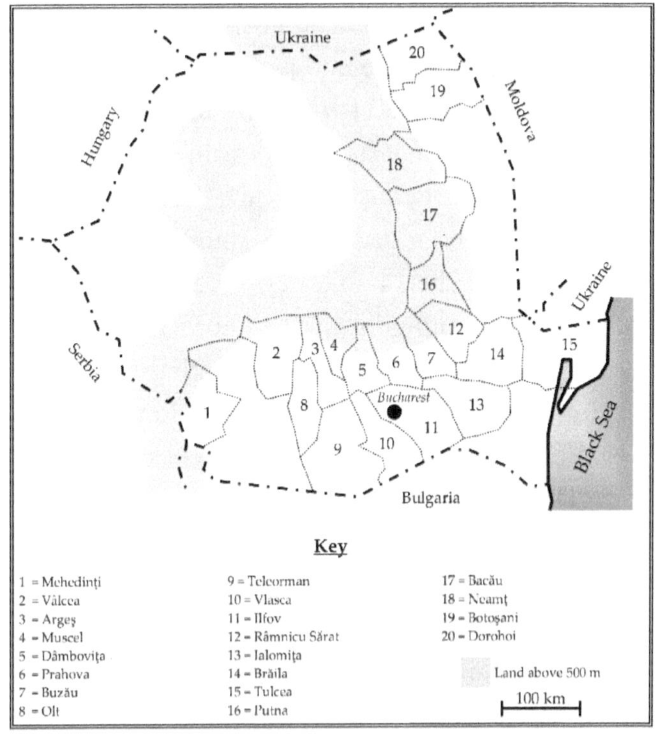

Sketch-map of Old Romania

To give some idea of where all the places are in Romania that Otescu frequently refers to, we present a sketch-map of the modern country, with the names of its surrounding eastern European nations. All the old Romanian counties mentioned are drawn on, and numerically keyed to their names. The capital Bucharest is also shown, along with the Black Sea coast, and the general highland area of the

Carpathian Mountains. The north-western parts of modern Romania were conquered and ruled by the Hungarians from the 11th century until the collapse of the Austro-Hungarian Empire in 1918. Despite the population being primarily Romanian, this region was only added to the country's area after Otescu wrote his book. The concentration of information coming from the more mountainous regions, away from the capital, is perhaps indicative of the older customs holding out longest there. Neamț, for instance, the single richest source of material used by Otescu, is an area almost exclusively composed of rugged highlands. Counties that contributed no information are not shown, and note that some of these counties no longer exist today. Those that do have often altered their boundaries significantly. The map and all the other illustrations in this book were prepared by Alastair McBeath. Those for the constellations were constructed following a combination of the diagrams and written information in Otescu's original publication.

Finally, we wish to thank especially Dan-George Uza of Astromix for all his many assistances in making possible this revised translation.

Andrei Dorian Gheorghe, Bucharest, Romania
Alastair McBeath, Morpeth, England
April 2024

ION OTESCU

*

ROMANIAN
STAR & SKY LORE

INTRODUCTION

The Romanian peasants consider the stars only after they have risen into the sky in the evening, at midnight and at dawn, between June to September, especially during July and August, the middle of their agricultural activity. Only the shepherds reckon the rest of the year; thus the stars they follow are called "shepherd's stars" by the other peasants.

*

What made me study the Romanian sky from a scientific point of view, and not leave it to other, more competent, people, from literary circles?

I think it is necessary to answer this question here.

From being a child, I knew many things about how the peasants divided the sky into constellations, and some of the beliefs and legends of the peasants about these star patterns. My parents were farmers. My father possessed estates in Ilfov, Vlaşca and Buzău counties, and lived with my mother among the peasants. With her investigative spirit, my mother was especially interested in the peasants' beliefs. As a cultured woman with a remarkable memory, she remembered very well every story, and had a special gift to re-tell them beautifully. When I was a child, she told me a lot of popular tales and legends, and a great many of them still remain in my mind.

She also showed me the sky, telling me many things about it, but as a child, I did not realise their importance. However, in part, I came to know what the peasants from Ilfov, Vlaşca and Buzău counties thought about the sky.

I imagined that there must be more than these peasants knew, from the population as a whole, and I waited for other men, as I said before, more competent than me, to collect and write about all these things. But seeing that my expectation still remains unfulfilled, I decided, after advice from some of my fellows, to begin this work myself. God grant that this beginning should be an impulse for those whose ability is richer than mine. To make this work as complete as possible, I intended to realize researches across the whole country, taking as my co-workers many teachers in rural areas. For this purpose, in 1896 I made a sky map, projected for the location of Bucharest. On it, I put all the astronomical constellations visible from Romania, in their classical Greco-Latin groupings, with detailed explanations for each constellation. Then I sent the map to all the teachers in Romania, with a special letter in which I requested them to ask the oldest peasants about their beliefs in the constellations, the sky, the Earth, Sun, Moon, eclipses, thunder, lightning, clouds, rain, etc., and anything else they wished to communicate to me. From the papers returned to me, plus my own researches, I have created what I think to be both an interesting and comprehensive work. It is true that many of the papers did not tell of anything significant, but in 39 of them, I found very important information. These answers came from the following counties: 1 each from Mehedinți, Muscel, Dâmbovița, Buzău, Râmnicu-Sărat, Brăila, Putna, Bacău, Botoșani, Tulcea; 2 each from Vâlcea, Olt, Ialomița, Dorohoi; 3 from Argeș; 5 from Prahova; 13 from Neamț.

Before entering into this material, I wish to declare something. I will present all as it is known to me, or as it was communicated to me. I wished to be consistent with myself and with my brave teacher co-workers. At the end of this work, I will present a list with their names and locations, and also some of the old peasants' names who made important contributions.

Finally, as I feel it to be necessary, I will reproduce antique ideas and beliefs, especially of the Greeks and Romans, about the constellations, heavenly bodies, celestial and meteorological phenomena. And I will try to show from which place, and how, some of these old beliefs came to us. Perhaps it is a daring risk for me to do so, but I considered it to be useful.

CHAPTER ONE:
CONSTELLATIONS

I: Ursa Major and II: Ursa Minor

Usually, these are called *Carul Mare* (the Great Chariot) and *Carul Mic* (the Little Chariot). The four stars which form the trapeziums in each [α, β, γ, δ Ursæ Majoris and β, γ, η, ζ Ursæ Minoris] are called *Roatele Carului* (the Chariot Wheels), and the remaining three [ε, ζ, η Ursæ Majoris and α, δ, ε Ursæ Minoris] are *Proțapul* (the Shaft), or *Tânjala Carului* (the Slow Chariot Shaft), or *Oiştea* (the Axle). In other instances, they are called *Ursul Mare* (the Great Bear) and *Ursul Mic* (the Little Bear) [16], the trapeziums being *Trupul Ursului* (the Bear's

Body), and the three separate stars *Coada Ursului* (the Bear's Tail). The Little Chariot is also called *Plugușorul* (the Tiny Plough) [14] or *Grapa* (the Harrow) [38].

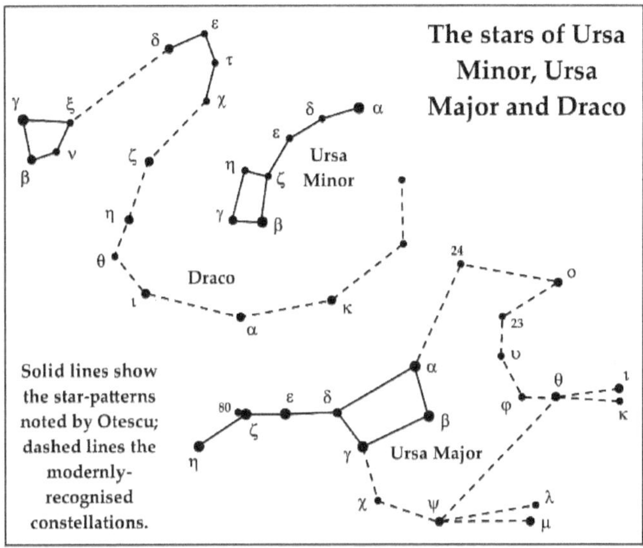

The peasants have grasped the important idea that the two Chariots are always to the North, and some of them use these as a direction guide at night [4].

Polaris [α Ursæ Minoris] is sometimes called *Împăratul* (the Emperor) [32], and at other times *Candela Cerului* (the Sky's Votive Light) [16], because it is a comparatively immobile star with little diurnal movement, and it is the most luminous star in that part of the sky.

Polaris is also sometimes called *Stâlpul* (the Pillar) [12, 40] or *Steajărul de la Arie* (the Stake of a Horse-

Threshing Area), which means the peasants have the understanding of Polaris' fixity, thinking that the sky revolves around this star, like the horses around the stake while threshing.

Other peasants called Polaris *Steaua Ciobanului* (the Shepherd's Star) [39], because it is a guide for shepherds in the night. However, the Shepherd's Star is also a name given to the planet Venus when it is *Luceafărul de Seară* (the Bright Evening Star) [29, 31, 40], as is also found in France. Finally, the pole star is also called *Țâgâră*, or *Țagâră* or *Țagâra* [regional dialect terms that can be translated as "Pointer" or "Clock Hand"; 29].

In the Great Chariot, near the second star in the Shaft [ζ Ursæ Majoris], there is a small star called *Alcor* or *Saidac* in astronomy [80 Ursæ Majoris]. Saidac is an Arabic word which means "trial", because it was used by the Arabs to determine the clarity of the atmosphere. The Romanian peasants call this star *Cărăușul* (the Carter) [39], who drives the oxen of the Chariot; other peasants call it *Cățelusa* (the Little Bitch), which accompanies the Chariot, and she sometimes has the name *Paloschița* [which is a Slavic name that simply means "Little Bitch"; 21]. However, some peasants see Cățelusa as the star near the Chariot Wheel, where the Shaft begins [which may be Alioth (ε Ursæ Majoris) or more likely the faint star near it (78 Ursæ Majoris;

not illustrated on page 25, it lies a little to Alioth's upper right on our diagram); 41].

Sometimes *Cărăuşul* is called *Ucigă-l Toaca* (He Who Will Be Killed By The Semantron) or *Ucigă-l Crucea* (He Who Will Be Killed By The Cross) [30] - who is the Devil - driving the oxen of the Chariot. Later, after we have looked at all the peasant constellations, we will find out more about this name in the Romanian Sky Myth. [A semantron is a percussion instrument made of wood or iron and struck with a mallet to make sounds like a chime or gong, and is used instead of bells among the Eastern Orthodox churches of southeastern Europe and neighbouring areas.]

Finally, the Carter can be called *Văcarul* (the Herdsman) [40], who drives *Şapte Boi* (the Seven Oxen), which are the seven stars of this constellation.

Usually, the Chariots are formed from the four Wheels and the Shaft; but some of the peasants imagine the four Wheels as the Chariot's body, and the next two stars are the Oxen of the Chariot. Polaris, the last star in the Little Chariot remains as the Sky's Votive Light or the Pillar, as we saw earlier, while the final star of the Great Chariot becomes *Ursul* (the Bear) or *Lupul* (the Wolf) [16, 18]. The oxen of the Great Chariot are afraid of whichever one he is, and try to run away from him. We will see how in the Romanian Sky Myth.

There is a beautiful legend about the Great Chariot, which runs as follows [37]:

A man called Pepelea had a wager with God that he could push the Great Chariot as far as the Milky Way. However, when Pepelea pushed the Great Chariot, God moved the Milky Way, and thus Pepelea lost the bet. For what can a man do against God?

This myth epitomizes the rotation of the Chariot and the Milky Way around the celestial Pole, as if the sky was a single piece.

Another myth explains the same thing and also accounts for why the Chariot has its shaft awry [24]:

The Great Chariot, which has four wheels and three pair of oxen [every star of the Shaft is here imagined as a couple of oxen; 23, 25], *climbed up a hill. At the top, God said to the Carter, "Up hill, up hill with God's help". But the Carter answered him, "With or without God, I am at the top". Then God, in order to humble the man, made the oxen go backwards, damaging a wheel. Ever since, the Chariot has always rounded the Pole backwards.*

Very important evidence for the Romano-Dacian origins of the Romanian star patterns is found in the way the peasants explain the use of these Chariots. They say that the Emperor Trajan put in the Great Chariot the slaves from Dacia that he had conquered, to carry them to Rome, and he set their chiefs in the Little Chariot. Paloşchiţa is then Trajan's bitch [5].

However, other peasants [26] said that Job from the Bible transported his treasures in the Great Chariot and then buried them. Afterwards, he climbed up into the sky in the Little Chariot; but this belief seems to be of Jewish origin.

The Romanian people also call the constellations Auriga and Perseus chariots. Auriga is *Carul lui Dumnezeu* (God's Chariot) [21], because at the world's end, Jesus Christ will transport all the righteous men to Heaven in this Chariot. Perseus is *Carul Dracului* (the Devil's Chariot) [33, 34, 35], because after Christ's Judgement, He will send all the sinful men to Hell in this second Chariot. "Heaven and Hell are both in the sky", the peasants say [21].

Obviously, the Great Chariot must be one of the first constellations imagined by man, because it is always visible in the same part of the sky, and its stars are luminous. The poet Virgil thinks in his "Georgics" that the first constellations developed would be the Great Chariot (which he calls Lycaon's Bear), the Pleiades and the Hyades.

In describing the appearance of Achilles' shield, Homer speaks of the Pleiades, Hyades, Orion and the

Bear or Chariot, which latter is the only one that does not bathe in the Ocean, that is, it does not set.

Additionally, Canis Major with the star Sirius, Taurus with the star Aldebaran, and Boötes with the star Arcturus, must be of this same early date.

In all probability, the name "Chariot" predates the term "Ursa Major", as being the most popular one with all peoples at all times, and tales were invented to try to explain the name of the Great Bear that only appeared later.

In mythology, Ursa Major is the nymph *Callisto*, Lycaon's daughter, King of Arcadia, and Ursa Minor is her dog, called *Cynosure* by the Phoenicians, which means "the dog's tail". Here is the relevant myth:

The sky god Zeus, having Hera (Juno) as his wife, but loving many other women too, seduced Callisto, who became Arcas' mother. Her protectress, the goddess Artemis (Diana) banished Callisto for this sin; but Zeus honoured Callisto and Arcas, lifting them up into the sky. Callisto took her dog as a companion, but the jealous goddess Hera transformed Callisto and her dog into two bears, and Arcas became Boötes.

In his "Metamorphoses", the poet Ovid considered that Zeus transformed Callisto into a bear on Earth, trying to hide the truth from Hera. One day, Arcas was out hunting, and thought his mother was really a bear. He tried to kill her, but Zeus appeared in time to save her, and set them both in the sky, where Arcas became Arctophylax, that is, the guardian of the bear. Then Hera, observing the stars that comprised Ursa Major and Ursa Minor, asked the sea gods to always leave them in the sky, and never to let them sink into the Ocean.

Later, the seven stars of Ursa Major were looked on as seven oxen grazing the sky's fields. Thus, this constellation became Septem Triones for the Romans, and its guardian became the bear-watchman, Boötes.

Ursa Minor seems to be a constellation developed after the Great Chariot. In the 6th century B.C.E., Thales called it "the Little Bear", and some three centuries after Thales, Aratus observed that the Little Bear was also called the Little Chariot because of its resemblance to the Great Chariot.

In other legends, the Little Bear was seen as Arcas, transformed into a bear, like his mother.

There is no constellation more universal than Ursa Major, because no other can have impressed the early observers so much.

The ancient Greeks called Ursa Major *Helice*, because of its turning around the Pole, to which it was nearer some 3,000-4,000 years ago. Later, it became called Ursa, perhaps because the bear was the sole animal known by the ancients in the polar regions. [Around 3000 B.C.E., the Pole Star was the star α Draconis, far closer to Ursa Major then than Polaris is now.]

The Gauls called it *Wild Boar*, the French knew it as *David's Chariot* or *The Pan*, and the last western bards from the Middle Ages sang of it as *Arthur's Chariot*. The Egyptians called it *Hippopotamus* (Horus-Apollo).

Athanasius Kircher, the famous German Jesuit mathematician, physicist, orientalist and philologist (1602-1680), called the trapezium of four stars *Lazarus' Coffin*, and the three others *Mary, Martha and Magdalena*. This name must have had an oriental influence, because the Arabs also called these stars *The Coffin and the Family Which Follows It*. Schiller, wanting to Christianize the sky, called it *Saint Peter's Boat*. The Chinese called it *Pih-Tou*, a kind of measuring instrument for wheat, like a half bushel, or an old cup, the four trapezium stars being the *Half Bushel* or *Cup*, and the three others *The Handle*, but later they called it *Ti-Tshe*, that is "the Sovereign's Chariot".

From "Bear", "Arctos" in Greek, and from "Septem Triones", the Northern Pole was called *Arctic*, or *Septentrional*, which means Ursa Major was considered the chief constellation of this circumpolar region from early times.

Ursa Minor was also called "Phoenician" because the Phoenicians were the first to remark its fixity in the northern sky, and used it as a guide on the sea, a fact which assured their supremacy over the Mediterranean trade routes for 1,000 years. Around 3,000 years ago, the Pole Star was the Beta star of Ursa Minor, the guide for the Phoenicians. This star was called Kochab later, from the Arabic name Kochab-al-Shemali, that is, "the Northern Star".

The Chinese impression was that the Pole Star was one that dominated the sky; thus they called it "The Great Sovereign of the Sky", just as the Romanians called it "The Emperor". The other stars of Ursa Minor were given names based on the Chinese Imperial Court - the Sovereign Star, the Imperial Prince, etc. - and we saw earlier that Ursa Major was their Sovereign's Chariot.

Finally, I observe that the star Alcor was considered by the Latins to be *The Carter* too, like the Romanians, but imagined as a child. Afterwards it became *The Colt*, who always stays near his mother.

III: Draco

This is normally called *Balaurul* (the Dragon, or Golden Serpent) [5, 39] but it is not seen as a darting snake; it is considered to be more like the *zmeu* from the popular ballads. Its two trapeziums are called *Încolăcitura* or *Colacul* (the Coil or the Twist), in its middle [δ, ε, τ, χ Draconis], and *Capul Balaurului* (the Balaur's Head) at one end [β, γ, ξ, ν Draconis; 39].

The zmeu from the ballads is a being from *tărâmul celălalt* [literally "other space", perhaps best thought of nowadays as meaning "another dimension", as somewhere beyond the normal world]. He is a kind of man, but unlike others. He has skin like a snake's hide, and he has a very long tail, giving him much the appearance of a serpent. He can fly in the air and is sometimes nicknamed *Zburătorul* - "the Flying Incubus". When he wishes, the zmeu can take off his hide, especially when he is alone with his wife, normally an earthly girl.

Credulous to stupid, the zmei are hoaxed by waggish men in many tales, but they are famous for their inconceivable strength, which makes them invincible and dreadful. Only the brave Feţi-Frumoşi [Beautiful Boys (plural); Făt-Frumos, "Beautiful Boy" or "Boy Beautiful", when singular] are able to fight with, and even defeat, them, but the combat is very difficult, and they often have to receive miraculous assistance, such as the raven who sprinkled one Făt-Frumos with water. Enervated and refreshed by this, the hero finally felled the zmeu.

Here is a tale about the constellation Balaurul from the Cătunul Mănăstirea Bistriţa village in the Neamţ County of Romania [25]:

The Dragon is called Balaurul. When the great rains and storms are unleashed, the Balaur revives and gambols among the clouds. When the clouds are thin, the Balaur can fall to the Earth. In olden times, a Balaur fell on a place called Balaur Hill, and he lay there for a long time, slowly rotting away. His length and thickness were very great, his skeleton was gigantic, and one of his ribs was two royal palms [approximately 55 cm] *in breadth.*

In astronomy, the Dragon from this constellation is the mythological serpent, the guardian of the golden apple tree in the Garden of the Hesperides. This monster was killed by Heracles, and the two were placed in the sky, Heracles being imagined with his right knee on the Earth and his left foot on the Dragon's head, but this explanation seems only to have been thought up later.

For the ancient Greeks, Apollo was frequently mixed up with Heracles, both of them being solar gods, Apollo generally representing the sunbeams, and Heracles the beams of the rising sun, so Heracles was practically a variant of Apollo. In mythology, Apollo killed the terrible dragon Python with his arrows, and at Apollo's birth, his father Zeus, the sky god, adorned him with a golden crown, gave him a lyre and a chariot drawn by swans, and sent him to

Delphi near the Castalian Fountain, the fountain of poetry and prophecy, to herald the oracles. Considering that Apollo was later replaced by Heracles, this myth would account for the appearance in the sky of the constellations Hercules, Draco, Sagitta, Lyra, Corona, the Chariot, Cygnus and the Fountain [γ Cygni] (though this may also have been an early name for Pegasus), which are some of the constellations around the Pole. In the same region are also the two Ursæ and Boötes, constellations connected to another son of Zeus, Arcas, as we saw earlier under Ursa Major and Ursa Minor. As we shall discover later, the constellation Delphinus can be connected to Apollo too, while the constellations of Aquila and Auriga (with the star Capella) are linked directly to Zeus, so all of this region near the pole would have been the celestial territory of the sky god. I have not yet mentioned Perseus (another son of Zeus) and the constellations Perseus, Andromeda, Cepheus, Cassiopeia and Pegasus, all of them connected with Andromeda's myth, because it seems these constellations were created later, when Apollo was already mixed up with Heracles.

It is true that I have not seen these things written anywhere, but it seems to me that they appear written in the sky!

IV: Hercules

Generally, the Romanians call this constellation *Omul* (the Man) [16], and in the people's imagination he is a lofty hero, a Făt-Frumos, who was the Hercules of Romanian myths, as we will see in the Romanian Sky Myth, in which the Man fights with the Devil, and worsts him.

The mythological aspects of Hercules I discussed when I described the constellation Draco.

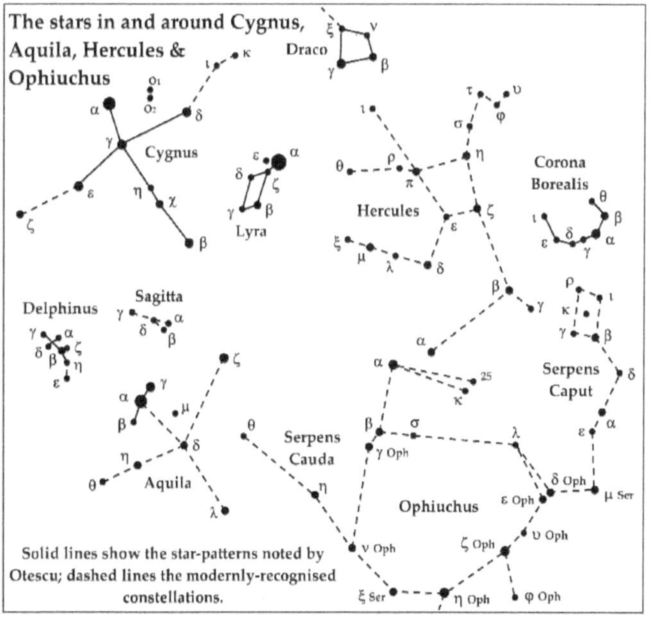

V: Lyra

This is called *Ciobanul cu Oile* (the Shepherd with His Sheep) [11, 36], the star Vega [α Lyræ] being the

Shepherd, and the four stars which seem to follow him, forming a quadrilateral, being the Sheep [β, γ, δ, ζ Lyræ]. A popular saying is: "In the days of Lent or Advent, people can eat cheese only after the Shepherd's rising" [36]. "The Shepherd is seen in the sky until sunrise", say the peasants [36]. The star Vega is also called *Luceafărul cel Mare de Miezul Nopții* (the Great Bright Star of Midnight) [12, 31, 32], or *Luceafărul cel Frumos* (the Beautiful Bright Star) [25], or *Regina Stelelor* (Queen of the Stars), because the stars have their leader, just like people on Earth. This is like the Lord's Prayer; "on earth as it is in heaven" [25].

The name Lyra in astronomy may be due to its shape. The quadrilateral formed by the four stars, *Oile Ciobanului* (the Shepherd's Sheep), can be imagined as a musical instrument, like the lyre, guitar or harp, with the star Vega being the fingerboard or neck of the instrument. The reason for this constellation being in the sky we saw in the earlier description of Draco, although it is also considered as Orpheus' lyre in mythology. The constellation Lyra was called *The Turtle* in the past, because ancient lyres were made from the shells of turtles. It is possible that its name came to be thus because of the form of the constellation, the star Vega imagined as the head of the Turtle, and the four stars its four feet.

Later, this lyre was perceived as a diving eagle, the name Vega deriving from the Arabic *Waki*, which means "The Swooping Eagle", in order to distinguish this constellation from that of *Al-Nasr Al-Tair*, that is "The Flying Eagle", which determined the name of the star Altair in the constellation Aquila [α Aquilæ].

VI: Cygnus

This star pattern is usually called *Crucea* (the Cross) [4, 6, 10, 20], or *Crucea Mare* (the Great Cross) [5, 25, 37], or *Crucea Miezului Nopții* (the Midnight Cross) [39].

In the popular imagination, this is the Cross on which Jesus Christ was crucified [5, 6, 17]. Thus when the peasants see it in the night sky, they cross themselves, saying that Christ showed Himself to them [32].

Other peasants call it *Fata Mare cu Cobilița* or *Fata Mare cu Coromâsla* (the Great Maiden with a Yoke, two terms for the same object) [11, 12, 21], the Girl being imagined with her head pointing towards Polaris.

Finally, other peasants see in this constellation only *Cobilița Ciobanului* (the Shepherd's Yoke), because "it rises and sets at the same time as the Shepherd" [Vega] [11, 36]. And they say that the Shepherd has two buckets on the yoke and another in his hand. The star in the middle of the Cross, called

Sadir in astronomy [γ Cygni], is named by the peasants *Fântâna* (the Fountain), or *Fântâna din Răscruci* (the Fountain of the Crossroads), and the Shepherd's Yoke is propped up by it.

The origin of the name for Cygnus as the Swan is not very clear. It may possibly be inspired by the brighter, white part of the Milky Way where Cygnus is, or the Swan may be the metamorphosis Zeus used to conquer Leda.

I gave my ideas concerning this constellation when I related the myth of Apollo's birth, under the constellation Draco.

Before finishing with the constellation of Cygnus, I must remark that the Romanian peasants see another Cross in the astronomical constellation Delphinus; another Great Maiden with a Yoke, *Fata de Împărat* (the Emperor's Daughter), formed by three stars from the constellation Aquila; and another Great Maiden, without a yoke, *Fata Mare din Horă* (the Great Maiden from the Hora or Ring Dance). This Maiden from the Hora is the star Gemma in the constellation Corona Borealis [α Coronæ Borealis], which is called *Hora* (the Ring Dance) by the peasants.

VII: Delphinus

As I just mentioned under Cygnus, this is also called *Crucea* (the Cross) [2, 18, 21, 31] or *Crucea*

Mică (the Little Cross) [36, 37]. The Arabs too called it Al-Salib, which means "the Cross".

However, its astronomical appearance is disputed. In its most important symbolic form, it is Apollo disguised as a dolphin attacking a ship which was carrying Cretans from Knossos, whom he afterwards made Priests in his Pythian sanctuary at Delphi.

So, this constellation's aspect can be equally imagined as a Fish or a Dolphin or as a Cross, "the handheld cross brought by the priests on the first day" [of the New Year], as a Romanian peasant described it to me.

VIII: Aquila

Aquila is called *Vulturul* (the Eagle) or *Vulturul Domnului* (God's Eagle). More generally, the peasants form a constellation from the star Altair and the two brighter stars to either side of it, which they call *Fata de Împarat cu Cobiliţa* (the Emperor's Daughter with a Yoke) [10, 14, 19, 23, 24, 38], where Altair is the girl and the other two stars are the Hooks of the Yoke, that support the pails in which the Emperor's Daughter carries water in order to refresh the souls of the sinful dead who are in Hell [18].

The ancients called this constellation "the Eagle" because of the two stars which lie laterally to either side of the star Altair, stars which very readily lead to imagining they are two wings that carry Altair aloft.

This was Zeus' Eagle, but it has become God's Eagle in Romania.

I have already told of the origins of Altair's name, in the description of the constellation Lyra.

IX: Cepheus

This is known as *Coasa* (the Scythe) [38]. The line between the stars α and γ forms *Coporâia* or *Coporâșca*, that is *Coada Coasei* (the Scythe's Tail). The star Beta alone is *Mânerul Coporâii* (the Tail Handle - of the Scythe); the stars μ, ζ and δ form *Fierul Coasei* (the Scythe's Iron Blade).

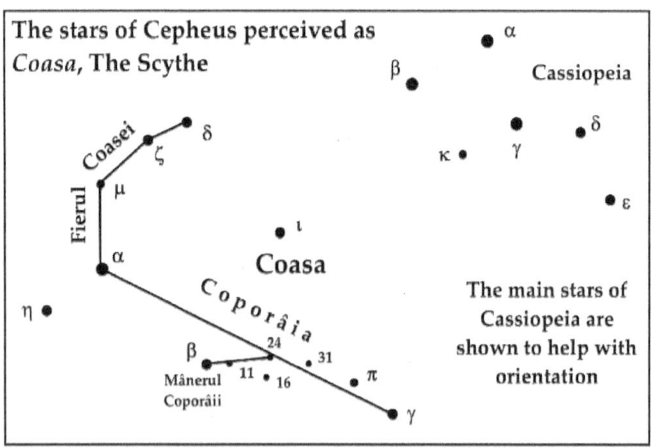

X: Perseus

Perseus is called *Căpățâna* (the Skull) [8, 16] or, as I said under the description of the Ursæ, *Carul Dracului* (the Devil's Chariot).

Some peasants call it *Barda* (the Broad Axe). Here, the line between the stars α (the brightest star in Perseus) and β (called Algol astronomically) forms *Coada Bardei* (the Broad Axe's Handle), while α on its own is *Muchia Bardei* (the Broad Axe's Top Edge). The part oriented towards Cassiopeia is *Gura Bardei* (the Broad Axe's Blade). A further alternative name for Perseus is *Toporul* (the Axe).

Algol is the astronomical name for the star which represents Medusa's Head, deriving from the Arabic *Al-ghul*, which means the Monster or the Devil, and on many star-charts, Perseus is called *the Devil's-Head Bearer*. This remembrance of Medusa's Head is also found in the Romanian peasants' names of the Skull or the Devil's Chariot.

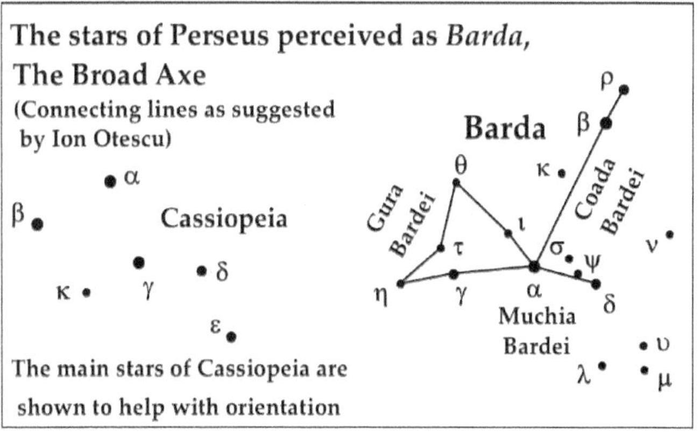

XI: Cassiopeia

This constellation is named *Scaunul lui Dumnezeu* (God's Chair) [39]. Its six principal stars produce the image of a chair. In astronomical mythology, Cassiopeia sits on a chair or a throne, although her husband Cepheus stands. Another peasant name for it is *Mănăstirea* (the Monastery) [37], because, of course, a monastery is also God's Seat or Chair.

XII: Andromeda

Andromeda is *Jgheabul Puțului* (the Drain of the Well), the Well itself being the astronomical constellation of Pegasus, discussed next.

XIII: Pegasus

Considering only the four stars which form the Great Square of Pegasus, this is called *Toaca* (the Semantron) [11], which was imagined as a percussion instrument hung beside the nearby Monastery (Cassiopeia). [These four stars are α, β, & γ Pegasi and α Andromedæ, the "top left" star when viewing the Square with north to the top.]

However, Pegasus and Andromeda together are called *Puțul cu Jgheab* (the Well with a Drain), the four stars of Pegasus' Square being the corners of the Well's edges, with Andromeda as the Drain.

It is possible that this name is a memory of Hippocrene, which means the Horse's Spring, a spring that appeared from a rock on Mount Helicon, after the winged horse Pegasus had struck it with his hoof. It may be the Pyrenean spring, from which Pegasus drank, or just the Castalian spring.

Pegasus is also called *Gavădul Mic* (the Little Stud-Farm) [16], *Gavădul Mare* (the Great Stud-Farm) being the constellation Leo, although Leo is more commonly known as *Calul* (the Horse) by the Romanian people.

The word *gavăd* here represents something analogous to stud. Mr. Candrea, a teacher colleague of mine at Mihai Viteazul High School, a man who dealt a lot with philological issues regarding Romanian words, told me that the word gavăd means something analogous to a pack of wolves or a woodchuck, and therefore flock, herd, stud.

Curiously, in the Chinese celestial sphere, this region is called *T'ien-Kiu*, that is, "the Celestial Stable", because these stars were in conjunction with the Sun in springtime, when the stables were cleaned in China, and sprinkled with the blood of a sacrificed horse, so this constellation could be of Chinese origin. This concept was probably the same for the Greeks, with Pegasus superposed later in order to complete Andromeda's myth. This idea was later assimilated by the Romans, and ultimately the Romanians, as a stud-farm.

The myth about the constellations Cepheus, Perseus, Cassiopeia, Andromeda and Pegasus is as follows:

Cassiopeia, Cepheus' wife (Cepheus was Ethiopia's king), considered herself more beautiful than the Nereids, the nymphs of the sea. They felt offended by the crazy pretensions of Cassiopeia, and asked Poseidon (Neptune) to revenge them. Submissively, Poseidon flooded the country, and sent to it a terrible marine monster, which ate the people and their herds. The Oracle of Ammon announced the way to deliverance: Andromeda must be sacrificed as prey for the monster. So, Cepheus chained his daughter to a stone near the sea, but Perseus, who was passing, fell in love with the beautiful maiden, and tried to save her. Riding Pegasus, Perseus showed Medusa's head to the monster, which died immediately. To remember these events, all these characters were placed in the sky, including Pegasus and Medusa's head (the star Algol).

XIV: Auriga

Usually, this is called *Vizitiul* (the Charioteer) [37], or *Trăsura* (the Carriage) [36], or *Surugiul* (the Coachman) [16, 36], while the star Capella [α Aurigæ] is *Capra* (the She-Goat), and the three stars near the She-Goat are *Iezii Caprei* (the She-Goat's Kids), which is also their astronomical name, the Kids. [The Kids are ε, ζ & η Aurigæ; see the diagram on page 51.]

However, the three Kid-stars are also considered a distinct constellation, called *Sfredelul Mic* or *Burghiul* (the Little Auger) [41] or *Sfredelul Pământului* (the Earth's Auger) [39].

Other names for the whole constellation of Auriga are *Țarcul* (the Goat Fold) [38], or *Carul lui Dumnezeu* (God's Chariot), as I said above under the two Ursæ.

The ancient Greeks saw in the constellation Auriga Erichthonius, the king of Athens, who invented the concept of harnessing horses to a chariot. In the "Almagest" of Claudius Ptolemæus, this constellation is called *Heniochus*, which simply means "Charioteer". The Charioteer was pictured from early times with a horse-whip and reins in his hands, which gave birth to the Latin name of *Tenens Habenas* (the Rein-Holder). He carried the She-Goat and her recently-born Kids on his left arm. The She-Goat is Amaltheia, who suckled Zeus.

It is difficult to say what associated the Charioteer with the She-Goat.

The Romanian people see three Kids, one in every star, but astronomical depictions show the two close stars as on the hips of the two Kids, and the third star is on the head of the Kid on the right.

XV: Boötes

Boötes is called *Văcarul* (the Herdsman). I discussed his origins under the Ursæ earlier. [We note here that one old Romanian name for the star Alcor, 80 Ursæ Majoris, is also the Herdsman.]

XVI: Corona Borealis

This is normally called *Hora* (the Ring Dance) [10, 18, 37], as I already said under the constellation Cygnus. The star Gemma [α Coronæ Borealis] is the *Fata Mare din Horă* (the Great Maiden in the Ring Dance), and the two smallest stars from the Ring [probably δ and ι Coronæ Borealis] are called *Lăutarii* (the Singers) [10].

In other cases, Corona Borealis is known as *Casa cu Ograda* (the House with a Courtyard) [38], Gemma being the House, and the remaining stars from the circle being the Courtyard. A further name is *Coliba* or *Cociorva* (the Hut) [32].

The myth concerning this constellation is:

Ariadne, King Minos' daughter, having been abandoned by Theseus on the sea-shore, was comforted there by Dionysus (Bacchus), who married her, offering her a wonderful crown. This crown, or only its precious stones, was later placed in the sky.

The Chinese call Corona "the Shell with Pearls"; curiously the Romans called Gemma "the Pearl" too, that is, the pearl of the crown.

XVII: Ophiuchus and Serpens

An early name for this star-pattern was *Şarpele* (the Serpent) [5, 38], in which is seen the snake that tempted Eve. This snake is the Devil, *Dracul* in Romanian, whose name's origin is from the Greek *drakon*, that is "serpent", a word which became *draco* for the Latin speakers, and ultimately *dracul* with the Romanians. More generally, the constellation is called *Calea Rătăciţilor* (the Road of the Lost) [33, 34, 35], a very apt name because it meanders in all directions. Onto this Road, the sinful people will stray, afraid of the second coming of Christ and His judgement.

XVIII: Pisces

This constellation is *Peştii* (the Fishes), or *Crapii* (the Carp) [16], but in either case, it represents the two fishes used by Christ to feed the five thousand [5].

In astronomical mythology, they are the fish disguises used by Venus (Aphrodite) and Cupid (Eros) to escape from Typhon's attack, or the two dolphins which carried Amphitrite to Poseidon (Neptune), to become his wife.

XIX: Aries

Aries is called *Berbecul* (the Ram) [16], but the peasants see only his horns.

The Romans called Aries Jupiter-Ammon, and the Greeks Chrysomallus, the ram who possessed the golden fleece, who was sacrificed to Zeus and then transported into the night sky. Aries is also the Apollo Corneus of the Dorians, and which in their expeditions went in the vanguard, just as a ram would lead a flock, or as the celestial Ram went before the other Zodiacal constellations. This latter idea seems to be the better one, knowing that the Ram was not one of the first constellations to be imagined in the sky. Aries was seen to contain the vernal equinox, and so preceded the rest of the Zodiacal constellations, like a ram before a flock. But it is also possible that the sight of this constellation, in which the horns of the ram rose somewhat brightly, may have aroused in the imagination a ram, or some other horned animal, and the ram was preferred for the above reasons.

XX: Taurus,
XXI: The Pleiades, and XXII: The Hyades

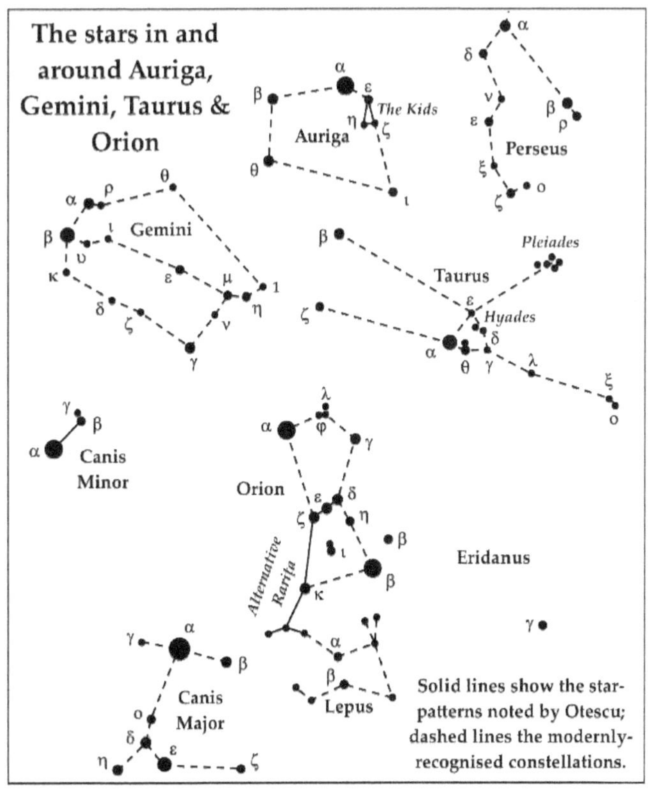

The people call Taurus *Taurul* (the Bull), or *Gonitorul* (the Mating Bull) [16], referring to the stellar triangle which seems to form the face, or the head, of a bull [the triangle formed from Aldebaran, α Tauri, and the nearby horizontal-"V"-shaped scattered star cluster of the Hyades, which lies to the right, and a little above Aldebaran, with north uppermost].

The star Aldebaran is considered a major star by the peasants, and they call it *Luceafărul Porcesc* (the Bright Star of the Pig) [3, 10], or *Porcarul* (the Swineherd) [27], because when it rises, the swine wake and grunt, signalling that day is approaching. It is also called *Deșteptătorul* (the Wakener) [3], because at its rising, the cockerels begin crowing at the day's start. Thus the peasants think that Aldebaran was put especially into the sky as a herald; the workers must wake, the ghosts must return to their graves, and the evil spirits must disappear from the earth when it appears.

The Pleiades [a much smaller, more condensed cluster of stars than the Hyades, known astronomically as Messier, or just M, 45] are formed into a separate constellation by the people, called *Cloșca cu Pui* (the Hatching Hen with Her Chicks), *Găinușa* [24, 27, 31, 37, 39] or *Găina* (the Hen) [4, 6, 10].

The Hatching Hen is the most important constellation for the peasants, and they always know where to look for it. In the winter months, the Hatching Hen is their night clock, and they measure time before day-break by this constellation's height in the sky [8, 15]. The name derives from the physical shape of the Pleiades, because the Hatching Hen walks first, followed by Her Chicks.

There are some special peasant beliefs about the Hatching Hen. He who sees it rising on Pentecost will have good luck [4]. He who sees it rising every day before Saint Peter's Day [June 29] will always be on-time for all events [37]. When the Hatching Hen appears in the sky at Drăgaică [June 24, Saint John the Baptist's Day], she has the power to make the corn grow [30].

Sometimes the Pleiades are called "the Stars of the Shepherd", for the shepherds observe when those stars set in the twilight in the spring, and then they say that the sheep may feed on grass [33, 34, 35]. Besides, the villagers also say: "When night falls and Găinușa is at the Semantron (that is, at the height at which the Sun is when priests announce the evening service at the church), then the people definitely know that the sheep are fully fed with grass" [26].

The Hyades also form a distinct constellation for the people, called *Vierii* (the Boars) [4], *Porcii* (the Swine) [26], or *Scroafa cu Purcei* (the Sow with Her

Piglets) [27; it seems likely this latter form would include Aldebaran as "the Sow"]. Only through confusion, some also call the Pleiades the Sow with Piglets. [In a footnote, Otescu then referred to the 1896 comments from a farmer in Neamţ county, Constantin Boboc, who explained that the group of seven stars was called Găinuşa (the Little Hen), and behind it was Porcarul cu Porcii (the Swineherd with Pigs). The two bright stars close by are the two brothers, obviously Castor and Pollux in Gemini. The same farmer believed that the planet Mars, which enters Gemini about every two years, was a third brother.]

The Romans too called the *Hyades Suculae* (Little Pigs), in accordance with the Greek *Siades*.

The Pleiades are very attractive, and were considered an important group from the earliest times. Cammile Flammarion wrote of them in his observational sky treasury "Les Étoiles" as follows:

Before the solar year was known, the first people regulated their calendar from the stars. The year began with the Pleiades' rising at dawn in the springtime, and the winter began with their rising at twilight in the autumn; the year was divided into two parts, and their appearance in November was saluted with a holiday to the dead, which is still kept today. The ancient Egyptians called November

Athor-Aye, *the Pleiades' Month, or Athor, and it was the same with the Chaldeans and the Jews. A similar division of the year was made by the Polynesians: one half of their year was called* Matarii i nia *(Pleiades Up), and the second half* Matarii i raro *(Pleiades Down).*

The Australian aborigines identically celebrate in November Mormodilek, *or "the Pleiades". The same habit is found in Mexico and Peru. The great pyramid at Gizeh is oriented to the four cardinal points (it probably served as an observatory) and has two interior galleries, the first directed to the north (to the Pole Star of four thousand years ago, α Draconis), the second pointing to the south at the Pleiades' highest elevation, which decided the year's beginning. For the ancient Greeks, Hesiod said agricultural works followed the Pleiades' appearance, while for the ancient Latins, they were called* Vergiliæ, *"the Spring Stars".*

The vernal equinoctial colure, which passes near the star α Andromedæ today, passed through the Pleiades four thousand years ago.

The Chinese Astronomical Annals retain an observation of this star cluster, made in 2357 B.C.E., as showing the vernal equinox.

In 570 B.C.E., Anaximander fixed their setting time at dawn, 29 days after the autumnal equinox. Some 3,000 years ago, the navigators awaited the moment of the Pleiades' rising in springtime before embarking upon their sea journeys, and the etymologists have decided that the name of the Pleiades derives from plein = *navigator.*

It is more probable that it derives simply from pleias = *a crowd, however.*

I continue to observe that today only six stars are seen well in the Pleiades. The ancients counted seven, and Ovid said that the seventh star ran away during the time of the Trojan Wars. However, those with strong eyesight can distinguish 10, or even 14, stars. In reality, there are more than 1,000.

The French peasants also call this ball of stars "the Hatching Hen with Her Chicks", like the Romanians, but they have another name for them too: "the Bunch of Grapes". Nine centuries ago, the Arabs called them *Al-sama ma banat'hi,* the Celestial Hatching Hen with Her Chicks. The star Aldebaran also gets its name from the Arabic *Al-debaran* (That Which Follows Closely) because it follows the Pleiades. Sometimes the Romans called it *Palilicium,* because the feast of Pales (the Palilia) was celebrated after this star's rising. The Arabs also called it "the Eye of the Bull", and the Jews, "the Eye of God".

The constellation Taurus played an important role in the ancient mythologies. To the Egyptians, it was the bull Apis; to the ancient Greeks, and especially the Cretans, it was the god Asterios, which symbolized Zeus as the bull abducting Europa. For the Romans, it was visualised as the same.

This constellation must be the oldest in the Zodiac. The others were created gradually, depending on their brightness and importance. In fact, Cammile Flammarion considered that observational astronomy appeared when the vernal equinoctial colure coincided with the star Aldebaran, in almost 3000 B.C.E. Later, the individual Zodiacal constellations were re-united into a single, theoretically-standardized belt of the sky, which was divided into twelve parts or signs, each of which was traversed by the Sun, month-by-month during the year.

The order for the moulding of these Zodiacal constellations seems to be: Taurus, Gemini, Leo, Virgo, Scorpius, Sagittarius, Pisces, Cancer, Aries, Capricornus, Aquarius and Libra.

XXIII: Gemini

This large constellation is reduced by the peasants to just the two principal stars, called Castor and Pollux [α and β Geminorum respectively] in astronomy. The Romanian peasants know them as *Fraţii* (the Brothers) [16, 27], or *Gemenii* (the Twin Brothers) [10]; although for them, they are not

Castor and Pollux, but Romulus and Remus, the founders of Rome [5].

Pollux is also called *Comoara* (the Treasure Chest), or *Comoara lui Iov* (Job's Treasure Chest) [26, 33, 34, 35], and there is a myth about this treasure, which will be related when I come on to discuss the Great Auger in Orion below. [Gemini is shown in the diagram on page 51.]

The ancient peoples called Gemini the *Dioscuri*, that is Zeus' sons. Sometimes they were known as Castor and Pollux, but at others they were Apollo and Hercules. The Dioscuri cult was spread throughout the whole of Greece and Italy. Castor and Pollux were the sons of Zeus, who had metamorphosed into a swan to lie with their mother Leda, and the twin brothers were born from the same egg. The glowing Aurora was born from another of Leda's eggs. Castor and Pollux were the gods who presided over hospitality and who quietened storms. Curiously, the aboriginal Tasmanians of Australia consider Castor and Pollux as their own Inventors of Fire.

XXIV: Cancer

This is called *Racul* (the Crayfish), and it has the following myth [5]:

When Christ was crucified, his torturers wanted to thrust four spikes into His body, two in His palms, a third through both His feet together, and a fourth, longer, one through His navel. But the Crayfish stole the longest spike, and took it far away, walking backwards. From that time, the Saviour blessed the Crayfish to be eaten on all the days of the year without restrictions. In order to keep alive the memory of its action, the Saviour decided that the Crayfish should walk backwards forever, and set it in the sky.

In astronomical mythology, the Crayfish was sent by Hera to bite Heracles when he fought with the Hydra, but the Crayfish was killed by Heracles, and placed in the sky by Hera.

XXV: Leo

Leo is rarely called *Leul* (the Lion). Instead, it is usually known as *Calul* (the Horse) [38]. It is also named *Gavădul Mare* (the Great Stud-Farm); *Gavădul Mic* (the Little Stud-Farm) is Pegasus, as I said earlier.

In mythology, Leo seems to be the apotheosis of the lion killed by Heracles in the Nemean Forest. In the Middle Ages, it was the Lion of Judah's tribe, or Daniel's lion from the den.

The star Regulus was known to the Greeks as *Basilicus*, that is "the Little King", because men born under this star's influence were considered to have a royal origin. The Arabs called Regulus *Al-maliki*, which means "the Royal Star". Copernicus latinized the "Little King" as Regulus, which is also called "the Lion's Heart".

XXVI: Virgo

The people call it *Fecioara* (the Virgin) [5, 6], who represents the Virgin Mary, the Saviour's Mother. The star Spica [α Virginis], being very white, indicates the Maiden's spotlessly clean heart.

The ancient Greeks called this constellation Parthenos (Athena), after their eternal virgin goddess. For the Romans, she was Minerva, goddess of wisdom. The Romanian people retain the idea of a wise maiden in their attribution to the Virgin Mary. The myth of Arachne also remained in the Romanian people's memory in this manner: the peasants think that if someone kills a spider, God's Mother forgives him nine sins, because the spider was cursed by God's Mother as praising itself for spinning better than her. Before this happened, the spider was a beautiful girl, and as a punishment, she was transformed into this revolting creature.

For the Romans, Virgo became in poetry Ceres, the goddess of corn sowing. She is also seen as Themis, the goddess of justice, because she has the scales of justice, that is the constellation Libra, at her feet. Alternatively, she is Astrea, daughter of Zeus and Themis, who was disgusted by men's crimes, and who went back into the heavens at the end of the Golden Age. In other cases, she is Diana from Ephesus; Isis from Egypt; Atargatis or Fortuna; Minerva or Athene; Erigone, Icarius' daughter; or Virgil's Sybil, who, carrying a branch in her hand, came down to the Underworld, which is beneath the northern hemisphere.

Sometimes a child was placed in her arms like the Egyptian Isis, which can be transformed easily into the Virgin Mary with Jesus in her arms.

I must observe here that the changed positions of Spica and Regulus were what led Hipparchus to discover precession.

XXVII: Libra

This constellation is *Balanța* (the Balance), or *Cântarul* (the Scales) [16], and Christ will weigh the facts of people's lives in them at His final judgement [5].

Mythologically, this balance is, as I commented under Virgo above, that of the goddess Themis, sometimes together with Astrea.

Some authors say that Libra was introduced into the Zodiac in Octavian Augustus' time, but it is known to have existed with the Egyptians three centuries before Augustus. Virgil attributes to it an origin that considers it as making the days and nights equal (at the equinox), but this constellation existed by that name before the equinox passed through it. Its name must have resulted from the equality of its two main stars. Sometimes it is called the Roman Scales, not necessarily because it was a Roman weighing instrument, but rather from the Arabic *rommana*, which means "weight".

Three centuries before Christ, Manetho wrote that the Scorpion's Claws had been changed into a Balance, so this constellation must be older than Manetho.

XXVIII: Scorpius

The peasants call this constellation *Scorpia*, that is, a female Scorpion [16]. She has "an eye of blood and long claws"; we will see why in the Romanian Sky Myth.

Obviously, the Scorpion was set into the sky by a people from hotter climates, where this creature is well-known, and from where the constellation climbs much further above the horizon. However, although the Romanian peasants did not know exactly what the scorpion was, they knew it was a fierce, wild creature, with darting claws, insatiable for blood. Certainly, this animal greatly impressed our Roman ancestors, remaining strongly in the Romanians' memory. "Thin and bad like a scorpia", say the Romanians about a dreadful woman.

In mythology, Scorpius represents the creature that stung Orion when he followed Artemis.

Its principal star is Antares [α Scorpii], which name means "rival to the planet Mars"; for the Greeks, Ares is the name of this same planet, the reddest body in the sky, so Antares and Mars vie with one another through their colours.

XXIX: Sagittarius

This is called *Arcașul* (the Archer) [16], who is a Roman warrior in popular Romanian belief [5].

The ancient Egyptian and Greek traditions tell of a kind of monster, half man, half horse, with the body, feet and tail of a horse, and the hands, neck and head of a man. These creatures were called centaurs, and were believed to be very strong, and very good archers. One of them, called Cheiron, was a physician and a teacher to the gods. He was considered the inventor of the heavenly sphere, that is, the first representation of the sky on a globe for studying cosmography. The ancients seriously believed these monsters existed: the naturalist Pliny said that in his time, a centaur preserved in honey was exhibited in

Rome. These things indicate that in those times there existed many speculative beliefs too.

These monsters could not have existed, because it would have been impossible for a whole race to disappear without trace. Thus we have a legend about centaurs, which was probably born when man began to ride horses. The first horsemen hunters, seen galloping and following game, using arrows and spears, would obviously have impressed those who had not yet developed the idea of horse-riding, and made them think the horsemen were unique beings.

In old depictions, the archer of this constellation is represented as a centaur with a bent, strung bow, whose arrow is aimed at the Scorpion. This centaur is thought to be Cheiron; for instance, Arago called Sagittarius "the Archer, or Cheiron". Another variant holds that this archer is not a centaur, but Crotus, a famous hunter who lived on Mount Helicon, near the Muses.

XXX: Capricornus

Capricornus is known as *Ţapul* (the He-Goat) [16], or *Cornul Caprei* (the Horn of the She-Goat) [5]. A popular belief is that when lightning appears from the Horn of the She-Goat, then rain is sure to follow [5].

The astronomical constellation is portrayed as a mythological monster, half he-goat, half fish, with the head and front feet of a he-goat, and the rest of his body and tail that of a fish.

He represents the form that Pan metamorphosed into, when he leapt into the waters of the Nile to escape from the giant Typhon.

XXXI: Aquarius

This is *Vărsătorul* (He Who Pours Out the Water).

Astronomically, this constellation represents Ganymede. In Homer's tradition, the gods hoisted Ganymede into the sky, and transported him to Olympus, to delight the eyes of the gods, and cheer their hearts, because of his beauty. He became the cup-bearer who poured nectar for Zeus and the other gods. In other traditions, Zeus disguised himself as an eagle, and abducted the beautiful Ganymede from the fields of Troy, and placed him in the sky. This explains why the constellation Aquila is above and near Aquarius.

*

*   *

In concluding here the series of the Zodiacal constellations, I observe that the people call these constellations *Zodii* (Signs), imagining that they are beings with the appearance suggested by their own names. Every man has a destiny depending upon the Sign under which he was born [22].

To complete the history of these constellations, I shall quote from "Les Étoiles" by Cammile Flammarion:

An interesting astronomical point is that of the Zodiac's antiquity...

Concerning the antiquity of the Zodiac, several authors have considered it to be formed 15,000-22,000 years ago. In directly studying the sky and its history, however, we saw that the Zodiacal zone was imagined after the creation of the constellations, and the constellations had varied successive origin times. Firstly, the Moon's path was observed, and this circumference of the sky was divided into 28 parts, each representing the Moon's house on every night of the month. Afterwards, it was recognised that the planets walk along the same belt of sky, and the Sun, in his apparent yearly travels, follows the same path... Thus it was possible to create a zone measuring 15 degrees in breadth among the pre-existent constellations that the Sun's way, or the ecliptic, traverses.

It is certain that in the time of Homer and Hesiod there were just a small number of named constellations, and that Strabo was careful to point this out to prevent his contemporaries from accusing him of ignorance. Eudemus of Rhodes, Aristotle's pupil, attributes the introduction of the zodiacal zone to the Greek sphere to Oenopides of Chios, a contemporary of Anaxagoras. Only towards the sixth century before our era, did the zodiac receive the names that have been preserved for us; yet Libra was not detached from Scorpius until the third century before our era.

We saw that the Pleiades played the principal role in the first calendar's creation. Since then, many thousands of years have passed since the Zodiac's route was traced by the courses of the Moon and planets; but it is not more than 3,000 years since our actual Zodiac was established...

Let us note that Aldebaran in Taurus, Antares in Scorpius, Regulus in Leo and Fomalhaut in Piscis Austrinus are almost at right-angles to one another, and divide the sky into four equal parts. These four stars, bright and remarkable, sometimes called royal stars, were venerated by the Persians around 2500 B.C.E. as the sky's four guardians. Then, Aldebaran, or the Bull's Eye, was at the vernal equinox as the guardian of the east; Antares, or the Scorpion's Heart, was at the autumnal equinox and was the guardian of the west; Regulus, the Lion's Heart, was at a little distance from the summer solstice point; Fomalhaut was at a short distance from the winter solstice point. These last two indicated south and north for the Persians. No doubt the Chou-King, which recorded China's early history, intended these stars in reporting that the Emperor Yao in 2357 B.C.E. ordered the astronomers Hi and Ho to observe the star Niao of spring, the star Ho of summer, the star Hiu of autumn, and the star Mao of winter, verifying at the same time the Sun's shadow. [The Chou-King is now more commonly known as the Shujing, the "Book of Historical Documents".]

XXXII, XXXIII, XXXIV, and XXXV: Orion

This is a constellation to which the peasants accorded great importance, but the stars composing it are grouped differently to what we find in astronomy.

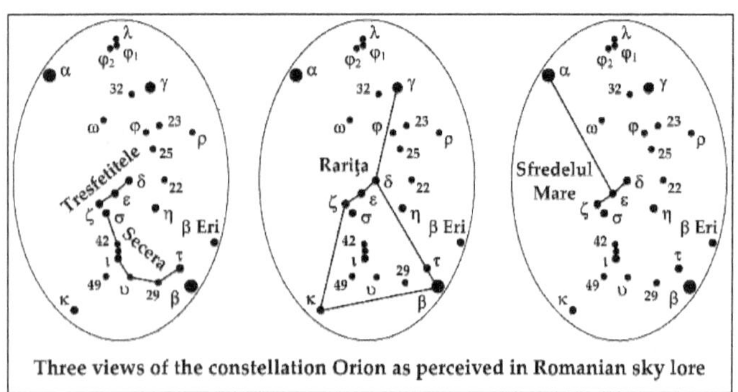

Three views of the constellation Orion as perceived in Romanian sky lore

The three stars from Orion's Belt [δ, ε and ζ Orionis] are called by the peasants *Tresfetitele* [21] or *Trifetiţile*, which means "the Three Saints", Vasile (St. Basil), Grigore (St. Gregory) and Ioan (St. John), or *Cei Trei Crai* (the Three Kings), or *Craii de la Răsărit* (the Kings from the East), who are the three wise men who came bringing gifts to Christ, or the three kings Saul, David and Solomon. They are also called *Toiege* (the Staffs) [1], or *Cingătoarea Regelui* (the King's Belt) [19], names which remind us of Aaron's Staff and Orion's Belt respectively, which are other astronomical variants.

Tresfetitele plus the stars Rigel, Saiph and Bellatrix [β, κ and γ Orionis] form the constellation *Rariţa* (the Little Plough) [4, 6], or *Rariţele* (the Little Ploughs) [9, 10, 18, 19], which announces the end of the night for the peasants: "Wake you up, for Rariţele has risen!" is an expression used before the corn harvest.

Others [38], however, form Rariţa from the Tresfetitele, with the star at the opposite lower corner of Orion to the star Rigel [κ Orionis], and three stars from the constellation Lepus, which come next to it [θ, η and ζ Leporis], forming an arc almost perpendicular to the line of the Tresfetitele [see the diagram on page 51 above].

Other peasants [6, 9, 10, 31] reduce the whole Little Plough to just Tresfetitele, or name it *Plugul* (the Plough) [38] or *Grebla* (the Rake) [19].

Between Rigel and Tresfetitele, there is an arc of seven stars. These stars form *Secera* (the Sickle) [33, 34, 35].

Tresfetitele and the star Betelgeuse [α Orionis] form the constellation *Sfredelul Mare* [6, 33, 34], sometimes *Spițelnicul* or *Spițelnicul Mare* (the Great Auger), which has as its point the star Betelgeuse, while Tresfetitele forms its handle. The Great Auger is directed with its point towards the Treasure Chest, that is the star Pollux in Gemini. Here is what the peasants from Neamț and Buzău counties think of it [26, 29, 33, 34, 35, 42]:

At the world's end, the Great Auger will approach the Treasure Chest, and pierce it. Then, all the money of the Treasure will pour out upon the Earth, but in exchange, all the water will disappear. Suffering from thirst, the people will run to the gold and silver fallen from the Treasure Chest, thinking it is water, but they will be disappointed. Then the Antichrist *(or Anti Hârs, or Anti Hârț) will appear, offering a cask of water and a chariot filled with bread for those who will submit to him. But those who trust him, drink his water and eat his bread will become even more filled with thirst and hunger. Those who refuse, and keep their belief in God, will prefer to die. Then*

Christ will appear to them, offering them a piece of wafer and a glass of wine, which will quench their hunger and thirst. This will be the Time of the Afterlife, and Christ will begin the Judgement of the Afterlife, or the Last Judgement.

It is curious how sometimes fantasy is in accordance with science.

Obviously, the peasants knew thirst many times, especially while working in the fields in summer under the Sun's scorching heat, and realized how terrible would be water's absence from the Earth. However, the leap from this, to the belief that in the future the Earth will be deprived of water is really something! Science proves that the Earth as it ages has less and less water, not because the water really disappears, but because as it percolates through the Earth's crust, water hydrates the subterranean rocks, and becomes part of their composition. This is a natural process on every planet, and we have a living example in the planet Mars, the Earth's smaller and older neighbour, which cooled faster than our planet. Another example is the Moon, which, although younger than the Earth, aged more rapidly because of its smaller dimensions. Here, however, opinions are divided, and there are certainly other causes for producing an absence of liquid water on a planet. In any eventuality, the people before science arrived foresaw this aspect through their fantasy.

In a paper from Ursați, in Neamț County, there is another myth:

The peasants tell the following about the constellations of the Great Auger and the Great Chariot. In olden times, a man searched for Job's Treasure Chest in the sky, and, discovering it, thrust the Auger into it; both Auger and Treasure Chest are seen today near the Slaves' Road [the Milky Way]. *Then he returned home, to collect his Ox-Chariot to carry the Treasure in. And where he was going with that Chariot being the curved road* [of the Milky Way], *is why you see the wooden control rod directing the steering on the right side* [where the Carter star, Alcor, is located alongside the three Oxen stars. In popular Romanian, the "control rod" is called a *proțap*]. *But in the future, when the man returns to put the Treasure in his Chariot, then the world will change. God will drain all the waters from the Earth, and down will come a rain of silver money. The people, discovering money and not water, will burden themselves with the coins. But Saint Peter will appear in their path with a basket of wafers and a glass of water, and will offer them to the people who were not so greedy for money, but will offer nothing to those who were too greedy. The first group will be satiated forever, but the second will remain thirsty and hungry.*

The constellation Orion has always occasioned many legends.

Orion was imagined as a giant with golden armour and a bright sword, very strong, and walking with great strides on his celestial way. With his Great Dog (the constellation Canis Major) by him, he follows the other stars, especially the Pleiades. In the summer time, when he rises at daybreak together with Aurora, he was considered to be Aurora's lover.

The sailors considered him to be a son of Poseidon (Neptune), the sea god. When they saw the bright constellation Orion appearing or disappearing over the horizon as their ship rode the waves, they said that the giant was stepping from one isle to another, making his way with his head in the sky and his feet in the sea.

Being invisible for part of the year, this aspect of the constellation Orion gave birth to the legends which tell of his death by Artemis' arrows, or because of the Scorpion, forced to sting him by this same Artemis.

The name of the star Betelgeuse is derived from the Arabic *Ibt-al-jauzah* (the Giant's Shoulder), and is also found as the variants *Btaljause* or *Betelgeux*. The star Rigel's name derives from the Arabic *Rijl-al-jauzah* (the Giant's Foot). Bellatrix has a Latin name, which means Warrior-Woman. In astrology, Bellatrix is the star of women orators.

XXXVI: Canis Major, and XXXVII: Canis Minor

They are called *Câinele* (the Dog) [16] or *Dulăul* (the Mastiff), and *Căţelul* (the Little Dog) respectively. The star Sirius [α Canis Majoris], the most brilliant star in the sky, is called by the peasants *Zorilă* (Dawnie) [10] or *Luceafărul din Zori* (the Bright Star of Dawn), because when it rises at dawn and the other stars all disappear, it seems to struggle against the dawn, until full daylight arrives. Sometimes the planet Venus is also called *Zorilă*, but usually the Romanian people accord to this planet only the names *Luceafărul de Seara* (the Bright Evening Star) and *Luceafărul de Dimineaţa* (the Bright Morning Star). However, as I said before under the Ursæ, the Bright Evening Star is also called the Shepherd's Star. The real situation is that the star Sirius appears at dawn in August, in the midst of the peasants' agricultural works, while Venus is sometimes, and sometimes not, seen at dawn at this time.

Other peasants call Canis Minor a second Little Auger (like that in Auriga) [6, 10], where the star Procyon [α Canis Minoris] is the handle, and the constellation's second-brightest star [β Canis Minoris] is the point.

The line of Orion's belt passes south-east through Sirius and north-west through the star Aldebaran, and this makes the peasants believe that these two are *Luceferii Zorilor* (the Dawn's Great Stars), mixing them up. [Both Canis Major and Canis Minor are shown on the diagram on page 51.]

The importance of the constellation Canis Major, most especially the star Sirius, makes me again resort to Cammille Flammarion's "Les Étoiles":

To ask at what age the men of the earth first took notice of Sirius, would be to ask how long men have been on this planet, for the first human gaze, which rose to heaven, had to be struck by the superiority of this heavenly light, and, even from the origin of astronomical contemplations, the sun of the starry sky had to receive the acclamations of admiring thought and the testimonies of human veneration.

The name of this star derives from the Greek "Seir", which essentially means "too bright", and was once applied to the Sun. Thus the adjective "seirios" became an epithet for an ardent, bright star.

The Greek word "Seir" comes from the Sanskrit "Svar", which also means "too bright".

In Sanskrit, the Sun was called Surya.

From Svar, too bright, was derived Seir, and then Sirius.

From Varuna, vault or canopy, came the name Uranos, sky, which became Uranus.

From Dyaus, luminous air, was derived Theos, the great being, and then Zeus, Deus and Dieu.

From Zeus and Pater were derived Jupiter etc.

3285 years B.C.E., and a century and a half before the building of Kheops' pyramid, Sirius regulated the Egyptians' calendar: its rising just before the Sun coincided with the summer solstice, and the flooding of the River Nile began with the first day of the month of Pachon (the Month of the Floods). This bright star was called Sothis (from the verb "to beam") by the Egyptians. Its role was to announce the rising tide of the River Nile, and it was epitomized as a harbinger Dog.

Homer and Hesiod saw Sirius like their ancestors from Egypt, Chaldea, Asia and China. Hesiod recommended harvesting the vines when Orion and Sirius culminated.

That bright star once rose [in the north of Africa] on the morning of the 21st of June, and although it heralded the rise of the Nile, it was at the same time the signal of the arrival of great heat. Thus the name of the Dog, "Canis", and the name of the Dog Star, "Stella Canicula", became synonymous with the great heat of summer, and we see the Latin poets Virgil, Horatius and Manilius recommending keeping a distance from the towns, and changing residence to the rural areas, during the days of scorching heat.

The heat wave begins then, if we believe Theon of Alexandria, 20 days before the rising of Sirius and ends 20 days later, this period having the privilege of giving rabies to dogs and colds to men. Sirius no longer rises on June 21, but precisely at the end of August it begins to appear before sunrise. However, for two thousand years the astrological prescriptions preserved the old date of the hot days, and today it is still possible to read in our almanacs of each year, that they extend from July 3 to August 11, a period that now has no relation to the empire of Sirius.

There is here a strange metamorphosis of ideas and words. The great heat does not derive from the idea of warmth, but from the concept of the dog (Latin: "canis", "caninus", "canicula"; in English: "the dog days"); the name "arctic region" does not mean cold at all, but the region of the bear. Septentrion is not so-called because it shows the direction of north, but because it recalls the seven oxen of the northern constellation...

For the most part, the ancient authors epitomized in Sirius the whole constellation of Canis Major, but today this name applies only to the one star.

This Greater Dog is, as we saw earlier, the giant Orion's dog.

Concerning the Little Dog, we observe that the name of the star Procyon means "before the Dog", because the Egyptians very carefully awaited the rising of Sirius, the Dog. Procyon appeared before Sirius, heralding it. Finally, it became named the Lesser Dog, while Sirius remained the Greater Dog.

XXXVIII: Cetus

Cetus is called *Chitul* (the Whale), and for the peasants, it is the monster which held Jonah in its belly for three days [5]. They do not know exactly what Chitul is; they think it is just a big fish. In the Biblical legend of Jonah too, we see this confusion of the whale with a great fish, while the Whale from the astronomical constellation is not just a whale, but a marine monster, because the ancients did not know the whale, which lives far away in the northern polar seas. In the Middle Ages, this constellation was also called the marine monster and the sea fish.

XXXIX: Columba

This is known as *Porumbița* (the Turtle Dove) [5, 19, 20], and the peasants think she is the turtle dove from Noah's Ark, who came back with the olive branch in her beak, after the Flood [5]. It is remarkable that Columba is one of the newest constellations, but uniquely was also the same for the Romanian peasants. In the scientific literature, we find that this constellation was first imagined by the Portuguese navigators of the 15th and 16th centuries; it probably came to us through trade; or perhaps this constellation is not so new as we think!

*

*   *

In finishing the constellations, I observe that the peasants also have other constellations. For instance, *Corbul* (the Raven) is probably the astronomical Corvus, which the peasants think was also Noah's, like the Turtle Dove, but which did not return to the Ark. *Lupăria* or *Haita de Lupi* (the Wolf-Pack) is probably Leo Minor, the Little Lion.

There could indeed be many more, because, although I made every effort to obtain the most precise information, I cannot be sure that I did not make errors.

In a communication from Cracăul-Negru, Neamț [32], the constellations Ophiuchus and the Serpent are called the Lesser Evening Stars; but I have every reason to doubt the accuracy of the place indicated, and rather to suspect that the Lesser Evening Stars are the star Antares and the stars which form the Claws of the Scorpion [probably the stars of modern

Libra]. At least this is what a peasant from Cheia of Prahova showed me; but he was not sure either, for once pretending to have forgotten what he had shown me, and showing him the three Kings and Betelgeuse in Orion, and asking him if they were the Lesser Evening Stars, he answered that they are; and to my doubt he himself confessed to me that he was not sure of their place. And whatever other peasants I asked in various localities, they did not know how to show me the Lesser Evening Stars, although they said they had heard of them.

As an example, a notice from Botoșani in Moldavia, Northern Romania, states that the astronomical constellation Sagitta, *Săgeata* (the Arrow) in Romanian, is called *Rază* (the Beam), but this is not confirmed by information from other Romanian regions.

It is surprising that the peasants, who later put all the most widely-used agricultural tools in the sky, did not also put the hoe, which serves so much, especially when working with corn. And certainly they did not, because I asked in depth, and everyone I asked told me that no hoe is in the sky.

Could the hoe be a newer agricultural tool, or at least one that was introduced to us after the Romanian sky had already been completed with all its constituents, and the ideas about this sky had been fixed?

The nature of the work is made, for nobody can say the last word was said.

*

*　　*

We have thus found that the Romanian people formed the following constellations:

I - The constellation *Carul Mare* (the Great Chariot), with the star *Cărăuşul* (the Carter) or *Ucigă-l Toaca* (He Who Will Be Killed By The Semantron) or *Ucigă-l Crucea* (He Who Will Be Killed By The Cross), the bitch *Paloschiţa*, *Ursul* (the Bear) and *Lupul* (the Wolf). This constellation is also called *Ursul Mare* (the Great Bear) and *Şapte Boi* (the Seven Oxen), with the star *Văcarul* (the Herdsman).

II - The constellation *Carul Mic* (the Little Chariot), or *Ursul Mic* (the Little Bear), or *Pluguşorul* (the Tiny Plough), or *Grapa* (the Harrow), with the star *Stâlpul* (the Pillar), or *Candela Cerului* (the Sky's Votive Light), or *Împăratul* (the Emperor), or *Steaua Ciobanului* (the Shepherd's Star), or *Ţâgâră*, *Ţagâră* or *Ţagâra* (the Pointer or Clock Hand).

III - The constellation *Balaurul* or *Zmeul* (the Dragon).

IV - The constellation *Omul* (the Man).

V - The constellation *Ciobanul cu Oile* (the Shepherd with His Sheep), with the star *Ciobanul* (the Shepherd), which is also called *Luceafărul cel Mare de Miezul Nopţii* (the Great Star of Midnight), or *Luceafărul cel Frumos* (the Beautiful Bright Star), or *Regina Stelelor* (the Queen of the Stars).

VI - The constellation *Crucea Mare* (the Great Cross), or *Crucea Miezului Nopţii* (the Midnight Cross), or *Fata Mare cu Cobiliţa* (the Great Maiden with a Yoke), or *Cobiliţa Ciobanului* (the Shepherd's Yoke), with the star *Fântâna din Răscruci* (the Fountain of the Crossroads).

VII - The constellation *Crucea Mică* (the Little Cross).

VIII - The constellation *Vulturul* (the Eagle), or *Vulturul Domnului* (God's Eagle), or *Fata de Împărat cu Cobiliţa* (the Emperor's Daughter with a Yoke).

IX - The constellation *Coasa* (the Scythe).

X - The constellation *Căpăţâna* (the Skull), or *Carul Dracului* (the Devil's Chariot), or *Barda* (the Broad Axe), or Toporul (the Axe).

XI - The constellation *Scaunul lui Dumnezeu* (God's Chair), or *Mănăstirea* (the Monastery).

XII - The constellation *Jgheabul Puţului* (the Drain of the Well).

XIII - The constellation *Puţul* (the Well), or *Gavădul Mic* (the Little Stud-Farm). The Drain and the Well together are *Puţul cu Jgheab* (the Well with a Drain).

XIV - The constellation *Vizitiul* (the Charioteer), or *Trăsura* (the Carriage), or *Surugiul* (the Coachman), or *Carul lui Dumnezeu* (God's Chariot), *Ocolul* or *Țarcul* (the Goat-Fold), with the star *Capra* (the She-Goat), and the three stars *Iezii Caprei* (the She-Goat's Kids). These three stars are also considered a distinct constellation, called *Burghiul* (the Little Auger), or *Sfredelul Pământului* (the Earth's Auger).

XV - The constellation *Văcarul* (the Herdsman).

XVI - The constellation *Hora* (the Ring Dance), with the star *Fata Mare din Horă* (the Great Maiden in the Ring Dance), and the two stars *Lăutarii* (the Singers). Other names are *Casa cu Ogradă* (the House with a Courtyard), or *Coliba* or *Cociorva* (the Hut).

XVII - The constellation *Șarpele* (the Serpent), or Calea Rătăciților (the Road of the Lost).

XVIII - The constellation *Peștii* (the Fishes), or *Crapii* (the Carp).

XIX - The constellation *Berbecul* (the Ram).

XX - The constellation *Taurul* (the Bull), or *Gonitorul* (the Mating Bull), with *Luceafărul Porcesc* (the Bright Star of the Pig), or *Porcarul* (the Swineherd).

XXI - The constellation *Cloșca cu Pui* (the Hatching Hen with Her Chicks), or *Găina* (the Hen).

XXII - The constellation *Vierii* (the Boars), *Porcii*

(the Swine), or *Scroafa cu Purcei* (the Sow with Her Piglets).

XXIII - The constellation *Fraţii* (the Brothers), or *Gemenii* (the Twin Brothers), or *Romulus and Remus*, with the star *Comoara* (the Treasure Chest), or *Comoara lui Iov* (Job's Treasure Chest).

XXIV - The constellation *Racul* (the Crayfish).

XXV - The constellation *Calul* (the Horse), or *Leul* (the Lion), or *Gavădul Mare* (the Great Stud-Farm).

XXVI - The constellation *Fecioara* (the Virgin).

XXVII - The constellation *Balanţa* (the Balance), or *Cântarul* (the Scales).

XXVIII - The constellation *Scorpia* (the Scorpion), with the star *Ochiul Scorpiei* (the Scorpion's Eye).

XXIX - The constellation *Arcaşul* (the Archer).

XXX - The constellation *Ţapul* (the He-Goat), or *Cornul Caprei* (the Horn of the She-Goat).

XXXI - The constellation *Vărsătorul* (He Who Pours Out the Water).

XXXII - The constellation *Tresfetiţele* (the Three Saints), or *Trei Crai* (the Three Kings), or *Trei Crai de la Răsărit* (the Three Kings from the East), or *Toiege* (the Staffs), or *Cingătoarea Regelui* (the King's Belt).

XXXIII - The constellation *Rariţa* (the Little Plough), or *Plugul* (the Plough), or *Grebla* (the Rake).

XXXIV - The constellation *Secera* (the Sickle).

XXXV - The constellation *Sfredelul Mare* (the Great Auger).

XXXVI - The constellation *Câinele Mare* (the Great Dog), or *Dulăul* (the Mastiff), with the brilliant star *Zorilă* (Dawnie).

XXXVII - The constellation *Căţelul* (the Little Dog).

XXXVIII - The constellation *Chitul* (the Whale).

XXXIX - The constellation *Porumbiţa* (the Turtle Dove).

These groups of constellations are almost identical with the Classical sky, which can be explained through the Roman origin of the Romanians. Many star-patterns even retain their Classical names. The changes that were made in the groupings were: simplifying some, which were reduced only to their essential, more striking parts, as in Gemini and Taurus, or subdividing others, or otherwise grouping them, which is rarer, like Orion. But in general terms, the differences as groups are almost insignificant, or small and few. With the names, however, the differences are much greater and more numerous, because there are many classical names that are of pagan deities, as the Romanians call them, which the Romanians who became Christian could no longer fully accept; and thus, either maintaining the old designation, gave it a Christian explanation, as is the case with the Virgin; or interpreted the old name by replacing it with a suitable one, but not a proper name, and thereby removing the myth, as is the case with Hercules, who was called Man, but implying a strong man, also a kind of Hercules; or changed them entirely, keeping only the names of animals and the common ones, such as the Charioteer and the Herdsman, which could also remain independent of the myths. What is more, after receiving them in this way, new legends were created about them, in connection with the beliefs and spirit of the Romanians, of which the most important and complete, which I will give in Chapter Three, embraces almost the entire sky.

Religious aspects were essential in this evolution. The Romans had many gods who formed another, superior, world to the human one, but like humans, their gods were very sinful. Roman mythology was wonderfully represented by poets, painters and artists.

However, the Romanians, abandoning the old mythology because they had discovered a new religion, with an ideal morality and a sublime simplicity centred around God, Christ and Mary, regenerated their flights of fantasy, enlightened by the sweeter, purer, milder sun of Christianity.

It is a well-known fact that in the legions of Trajan's army, with which he conquered Dacia [now modern Romania; Trajan's conquest of Dacia was completed in 106 C.E.], there were very many Christians, so the Romanian peoples' origins are connected with the beginnings of Christianity here. Only in Dacia were Christianity's adepts able to more freely practice their religion, as they were vigorously persecuted throughout all the other provinces of the Roman Empire.

For Trajan's Christian soldiers and colonists, this new homeland was especially precious, and their gratitude to Trajan, because he had conquered the country, was all the greater. Trajan became like a father-figure for them, and they created those legends concerning Trajan and Dochia, and Făt-Frumos, in which it is seen that they compared this Emperor with the Sun. In fact, mythology is only a legend to explain the whole universe's origin, imagined by our ancestors, based on their observations of all the phenomena around them. Contact between the Mediterranean and local myths created transformed myths, sometimes with confusions and controversies between them. Obviously, mythology is a difficult, vast science, which must be carefully linked with history and philology.

Just like other peoples, the Romanians put their myths into the sky, composed from memories of their Roman origins (the two Chariots of Trajan,

Romulus and Remus, the Roman Archer, etc.), elements of their new religion, and pagan names from old Dacia. They filled-up the remaining gaps with agricultural items and people, which explains why they see in the sky things like the Cross, Monastery, Plough, Semantron, Scythe, Sickle, etc.

This agricultural apotheosis is not surprising, agriculture being the Romanian peasant's life. The later Romanians too were a people who very much honoured agriculture, but they could not put it in the sky, because their sky was full of myths borrowed from ancient Greece, and the Greeks were not farmers, because of their difficult soils. Instead, the Greeks were very good sailors, and their mythology refers to the sea, the clouds and the winds. Thus, the Greeks did not forget their sea, and the Romanians, original creators of a part of the sky's myths, did not forget their agriculture.

Finally, I observe that the majority of the constellations imagined by the Romanian peasants look like the objects they are supposed to be, a rare thing in astronomy.

CHAPTER TWO:
THE EARTH

The Romanian peasants do not have the idea that the Earth is a sphere isolated in space. For them, the Earth is a plane, "like the face of a table"; at its edges, the Earth meets the sky, and it is supported by various things from below [8].

Some of them think the Earth is propped up by several pillars (usually three), made of iron, or wood, or wax. These pillars are gnawed at by an old *zgripțuroaică*, and it is her activities which produce earthquakes. This zgripțuroaică - who is a kind of supernatural monster-woman - eats only corn and water from the people of the Earth. When women sift corn, the part that remains around the edges of their sieves is thrown away for zgripțuroaică. Similarly, when women go to the spring to fill their pails with water, the water they use to wash the pails with is thrown onto the ground to be drunk by zgripțuroaică. Thus, say the peasants, economical thrift is good, because any kind of waste can give an enemy advantages [9].

Other peasants set Judas Iscariot in the place of zgripțuroaică.

There are also peasants [38] who say the Earth is a substance created by God, and set by Him into water on the back of a big fish. When the fish moves, the Earth shakes [8, 9, 38]. As evidence for this, they affirm that wherever one goes on Earth, if one digs deeply enough, water will appear.

Some put the fish and the pillars together as supports for the Earth, but others have two fishes or two buffaloes [8], because they are also able to stay in the water.

Finally, other peasants [9] explain earthquakes as being produced when God looks down, "because the Earth knows that on it, many sins and wrongdoings happen, and it cannot look at God without shaking from fear."

I shall conclude this chapter by telling some legends.

The first explains the creation of the mountains and valleys:

God [9] created the Earth as a waste field, but He remarked that it was too large to fit under the sky. So, He sent the bee to ask the hedgehog for a solution. The hedgehog referred to his own rolled aspect.

"If the Earth cannot be fitted under the sky as it is too smooth everywhere, it must be tightened, rolled up, wrinkled with hills, mountains and valleys."

Then the bee told God of the hedgehog's opinion, and God finished the Earth off as we see it today, with hills, mountains and valleys. Also, since then we started eating bee honey. [Here, we have preferred a much-simplified version of this legend, as Otescu's original included many difficult-to-translate regional dialect terms and concepts. In the remainder of this chapter, we have reproduced some other short legends that were in Otescu's footnotes.]

The magazine "Şezătoarea", year I, page 232, offered another popular explanation:

Once upon a time, the fishes which support the Earth on their backs became angry and writhed about greatly, making the Earth shake terribly, and the water below became very stormy, which created the valleys, mountains and hills. At the same time, springs were opened in the Earth, which are the source of the rivers.

Also in the magazine "Şezătoarea", year III, page 25, is an explanation for the Earth's creation:

In the beginning, when the Earth was not yet created, and only the water was in place, God called the frog to Him, asking her to go to the ocean bottom to look for some earth. After a long time carrying out this order, the frog came back to God with a little earth in her mouth, telling Him that at the ocean bottom there was a lot of earth. So God commanded the waters to move back to the sides, and leave the Earth above. His order was executed, and from that time forth, the Earth has stayed above the waters. From then on too, God blessed the frog, and he who kills a frog commits a grave sin, so that on his death he will have boils all over his body, just like the frog [an idea that seems to conflate smoother-skinned frogs with toads]. *The weaving of the Earth* [a term used here instead of "making" or "creation"] *was on a Tuesday, and that is why it is considered a sin and not good to sew shirts, sheepskins, drawers, to start new houses, make pots and all kinds of other things on that day, because you will upset God.*

Another legend is this:

One day, God spat in his palm, and His spittle became the Earth. From that day onwards, the Earth stayed on God's palm.

And one last legend:

The Earth is supported by six wax pillars. Judas gnaws eternally at them, to sink the Earth down and destroy the world, but when he finishes one of them and runs to the next, the gnawed pillar springs up again. At Easter, when he sees the Christians with their red eggs, Judas becomes very nervous and loses his mind. [In Romanian Easter traditions, eggs are dyed red to symbolize the blood of Jesus Christ that flowed from the cross onto the eggs in the basket of the Virgin Mary.]

CHAPTER THREE:
THE SKY, STARS AND FALLING STARS

For the peasants, the sky is solidly material, with an edge at the place it meets the Earth. It represents the bed clothes covering God's feet [22]. However, the most cultured peasants recognize that we do not actually see the sky, simply its blueness, because it is so very far off [9, 22].

In their conception, they imagine the sky as being:
* a bridge between the celestial world above and the terrestrial world below [8, 37];
* a bridge like a thick iron arch which rests on the waters and upon several pillars. Near the waters, it has gates which are used by the angels to inform God about new things they have

found in the terrestrial world [38];

* another Earth, which can be used for ploughing and sowing, because it has fields with no edges (as we shall see in the Sky Myth) [9, 24];
* a skin which stays suspended in the air through God's power (as we find in David's Psalm, who said to God: "You who stretched the sky like a skin") [10];
* made of two pieces which can be united or opened along the joint indicated by the Milky Way.

Like the Earth, the sky is perceived as lived-in, but by God and His angels and saints, not by men [17, 21]. The souls of the righteous also go into the sky, to God, after their deaths. Hell too is placed in the sky by some, but further off. Other peasants imagine that Hell lies beneath the Earth, and that in it exist the fires of eternal torture.

Every night in the sky, the angels and saints carry out religious services, as priests do in the monasteries. When the angels use the semantron in the sky, the cockerels hear it, and begin to crow. The sky's semantron is also heard by good and faithful men on Earth.

On some nights, the sky is said to open [in a magical or supernatural sense, allowing more direct contact with beings and powers in the heavens], such as on St. Vasile's (Basil's) night [at New Year; 9],

Boboteaza (Epiphany) Night [January 6, when Christ was baptised], and Învierii (Resurrection) Night [the Orthodox Church's Easter], but this is noticed only by the good believers. At these times, believers can ask God for anything, riches or health, and God will grant them all they desire. When the sky is open, the cattle can speak, and their voices can be heard by these same people. However, other peasants think that the sky is open like this only on the night of Ovidenie, November 21 [The Great Entrance, this is the day when Joachim and Anna took Mary to the Temple and entrusted her to the priest Zacharias] and that its beauties can be seen only by "shepherds who have never seen a woman's eyes". The sky and the Earth at this time are supported by St. Simeon Stylites, who lived on a pillar.

As we saw earlier, the cockerel has an important role in this Christian myth, but not like that in the pagan legend where Electryon was changed into a cockerel as a punishment, as he failed to keep watch during the love-making between Ares (Marte, or Mars) and Aphrodite (Venera, or Venus). Instead, we have the cock heralding the celestial service; his final crow, just before daybreak, has the power to drive away evil spirits, which suddenly vanish. Ghosts that are surprised by the cock-crow remain fixed where they stand, which is why they always hurry away to their graves well before the cock sings.

The ancients considered only this final crow of the cock in their legend, which announces the sunrise, that is, Apollo's arrival; it was this which Electryon failed to warn Ares of in good time.

Some peasants [21] believe in the existence of nine skies or heavens, in the last of which is God's Chair or Throne, where God lives with His angels. They also think that in every one of these nine must exist another distinct Earth, lit by its own Sun.

The canopy of heaven is lit in the night by the Moon at times, and the celestial candles, called stars, which are re-ignited by the angels every evening, to provide light to the Earth each night [38]. Here, the people are very naïve!

The stars, as I said, are candles in the peasants' beliefs, and every man has his own star in the sky [9, 22]. Each man's heavenly candle is set alight at his birth, or near him after he dies. If that man is an emperor, then his star is great and brilliant, a luceafăr. If the man is a simple one, his star is small and faint. When the man dies, his star falls from the sky, and his candle is extinguished. And those who believe this, look to see where the star falls, because that is where the one who died must be.

As we see, the peasants do not realize what falling stars really are, but think they are like the usual stars. For the peasants, all of Creation is God's gift to man, who is His beloved child. But if every man has a star, and that star is connected to his life, why are the stars in the sky always the same? Why do the constellations remain identical? Are some men immortal? The people, of course, do not ask these questions. It is curious how little the people control their beliefs. How sometimes they can be very profound, but at others they can be so childish...

Such inconsistencies among the various beliefs, however, are present with all peoples at all times, and this is especially so in this particular instance. For example, the story regarding the Kings from the East, who, on seeing Christ's star in the sky, a new and very bright one, thought that a great emperor had been born, and went to pay homage to him; but this was a sighting of a comet, not a star.

Falling stars, candles that are extinguished while in flight leaving a brief trail of light, do not reach the Earth.

Thinking of fireballs [very bright "shooting stars" or meteors], the peasants add that there are other falling stars, which are round or long, and which can enter into men's houses, or fall to the Earth, or even land on men [3]. These are the flying balaurii (the Romanian dragons) or zmeii (balaurii in human form), which walk in the night to disfigure or kill lone men. This is because the zmeii are evil beings. Thus, these stars are also known as wandering or travelling stars.

When the zmeii come into men's houses, they search for boys and girls to torture, a belief which was magnificently described in the poem "Zburătorul" ("The Flying Incubus") by the incomparable Ion Heliade Rădulescu. [Published in 1943, this poem is especially famous for the line "dragon of light with a flaming tail", which expresses the image of a meteor.]

Of course, these zmeii have their own conflicts, and they can fight by hitting their heads together until they die. Their blood then falls to the Earth, becoming coagulated and blackened. The peasants think that if they burn this blood in their houses, the smoke will drive away their children's fears. Some teachers from Neamț County told me that this blood must be coal from the Earth, but I think it more likely to be ozokerite [a naturally occurring, waxy mineral found in areas where crude oil is present].

Travelling stars can be bewitched, and this witchcraft helps girls determine in which direction their future lover lies. This is possible because there are said to be men who can say some magic words to stop a star falling, and then change its direction, and where the star then starts, there is the chosen one for the girl.

In finishing the Romanian people's beliefs in the stars and the sky, I recall what my mother told me about the stars.

The sky is supported by four pillars, which are four

stars. I think they must be the four royal stars, Aldebaran, Antares, Regulus and Fomalhaut. In the ancient world, these were the sky's guardians, but for the Romanians, they are the supporters of the sky. It is wonderful how these two distinct ideas can be reconciled.

Certainly, even when called the Pillar, Polaris is not perceived as the sky's pillar, but more as a stake the sky revolves around.

My mother also told me about seven stars in the sky, stars which never stop moving; when they all come together in the same place, the Earth will be destroyed. Of course, these are the planets, the seven planets of the ancient world, because the people then did not separate the planets, Sun, Moon and stars. Perhaps the ancients observed that planetary conjunctions coincided with terrestrial disasters. Something similar was written in recent years concerning the conjunction between the giant planets Jupiter and Saturn.

The ancient Greeks lionised the sky in two phases: the starry night sky, Uranus, and the luminous day sky, Zeus (or Jove), which were the bases of their ideas of the sky.

For the Greeks, the primordial trinity of genesis was formed by Chaos, Gaia and Eros, in which Chaos was infinite space, full of darkness, Gaia was terrestrial material which became the Earth, and Eros was the attractive force of universal gravitation. From Chaos were divided two parts of the darkness, a masculine and a feminine principle, Erebus and Nix or Night respectively, which gave birth to Aether and Hymera, the region of light superior to the Earth's atmosphere, and the terrestrial atmospheric light respectively. Thus, as in the Biblical Genesis, light was the first-born.

Then Gaia alone, but influenced by Eros, gave birth to Uranus, the starry sky, followed by the mountains and abysses of the sea, Pontus. Uranus became Gaia's husband, and together they were the primitive immortal couple, the source of all life, previously celebrated by the Hindu Vedic poets. The last child born of this pairing was Cronus, time. Finally, Gaia was revolted by Uranus' cruelty (he had halted the work of Creation, and hurled his children into the abysses of the Earth), and pushed Cronus to maim Uranus in order that he might no longer be able to procreate. Then Cronus and Rhea, the flow of time, became the second couple of Creation. However, later, Cronus too tried to halt Creation, by swallowing his children. His last son, Zeus, defeated him, bound him in chains, and thus became leader of Creation. With Zeus, historical and mythological time began.

The Latin poets sometimes just said "Jove" for "the sky". For instance, in Horatius we find *sub Jove frigido* for "under the cold sky".

The Romanian people made the sky into the house for God alone. For the rest, their physical conception is almost identical with the ancients. Anaxagoras of Clasimene said that the skies were formed from stone. Anaximander, Anaximenes and Eudoxus of Cnidus considered they were formed of a solid material. Empedocles thought that the sky was made of air, transformed into crystal by fire, and that the stars were of fire that appeared from the Earth, like the emanations of inflammable gas, which, following on from Xenophon's addition, were kindled in the evening and extinguished in the morning.

Plato described the concave vaults which formed the sky.

Aristotle imagined that there were many skies made of clear crystal, so that they could be traversed by light. Each of these skies belonged to a separate, concentric sphere, with the Earth at their common centre, and each one supporting a single planet, such as Mercury, Mars, the Sun, etc., until sky number eight, which was the Firmament, that is, the sky of fixed stars. Beyond that was an infinite vacuum, although other authors admitted of the existence of a ninth sky, called the *Primum Mobile* [Prime Mover], which connected with the remainder of the skies through a common, uniform, movement.

The ancient Jews held similar beliefs. In the Talmud, it is written that the rabbi Barchana touched the point where the sky met the Earth, and hung his hat on the window of the sky. When he wished to retrieve his hat, however, he could no longer see it, because the sky had moved it away in its rotation. So the rabbi is waiting yet for the celestial sphere's revolution to bring back his hat, and he stands with his hand held out to catch it on its return. [The "rabbi Barchana" Otescu mentions here was actually the Talmudist Rabbah bar bar Hana, about whom a great many, sometimes fantastic, stories were recorded in the Talmud. This tale is related in Bava Batra 74a.6 of the Talmud, although there, it was actually Rabbah bar bar Hana's basket, not his hat, that he left in a window in the heavens, and he only had to wait until the following day to retrieve it.]

Romanian Sky Myth [16, 42]

After the world's creation, the Sky and the Earth were very close together. But Man, as he is insincere by nature, did not realize this divine generosity, that it was no small thing for man to have God around him, so that he could ask him for advice, like a parent, or when he had any need. Man's indifference was so pronounced that one day a woman threw a child's stained nappy into the sky - although fortunately, it did not actually touch the sky. God became very upset at this, and removed himself and the Sky far away from the Earth, so that we now say of something, "It is far away, like the sky".

The woman was the culprit, for the woman has "a long dress and a short mind" or is only "good to make trouble". But Man suffered because of God's absence, and so he decided to journey to the Sky, looking for the Creator.

He knew the road would be difficult and long, so the Man took with him his Great Chariot with four oxen, his Little Chariot, his Votive Light from the wall, the Great Cross of the Church, the Fountain of the Crossroads, his Broad Axe, his Great Auger, his Little Auger, his Sickle, his Scythe, his Great Plough, his Little Plough, his Mastiff from the sheepfold, his Little Dog from the courtyard, the Hatching Hen with her Chicks, the Sow with her Piglets, the Shepherd, the Herdsman, the Charioteer and Hora the Ring Dance from the village. Because he wanted to appear before God like a good Christian, he took with him the most necessary human beings and things, to provide him with cheery company and help in times of trouble. He also took seeds and wheat to till and sow in the sky's fields for his future food.

Then he set off, and he travelled for a long time, but in the middle of the Sky's road, "He Who Will Be Killed By The Cross", the Devil, stopped him.

"Where are you going?", asked the Devil.

"None of your business", replied the Man.

"Who are you looking for?", queried the Devil.

"Leave me alone!", shouted the Man.

"You are a petulant fellow!", retorted the Devil.

"That's not true! You are a scoundrel and an evil one!", was the Man's response.

Then the Devil, furious, pulled out from his bag the Dragon, the Violent Serpent, the Great Bear, the Damned Scorpion, the Angry Horse and the Skull, and threw them around the Man to scare him.

But the brave Man began to fight with the Devil, and their battle created a great storm beneath the Sky, called by us on Earth "the rabid wind", because every flying creature that reaches this wind becomes rabid immediately, and falls down dead; and whatever beast eats of it, also is made ill.

In the meantime, the strong Mastiff and the Little Dog attacked the Horse, so the Horse ran far away. The Shepherd struck the Dragon Balaurul with a yoke. The Herdsman banished the Serpent with the Ring Dance and the swirls of the snake's flight can still be seen in the sky today. The Charioteer crushed the Skull with the Axe. With the battle going so badly, the worried Scorpion, which had tried to seize the Man with her claws, burst in a moment, and blood gushed from her eye. Only the Oxen of the Great Chariot ran off scared of the Bear, when the frightened Great Bear became rigid with terror. But the Bear also froze with fear, when he saw how the Man had crushed the Devil, and the Devil was next to the Bear; and the Bear saw well that his turn would soon come, which is why the Bear stopped growling.

Now, we see these events in the sky. In the midst of all these creatures, people and things, the victorious Man appears large and grandiose, while the Devil is very small and huddled.

God made the Man to be king in the Sky, as on Earth. Even the Devil recognises the Man as his master!

But the Man still has a long way to go to reach God. He trusts in God's help when he asks Him something with a pure soul.

We know too the Man's road. It can be seen on clear nights without the Moon. Its name is the Milky Way. It became white because in the battle the Shepherd accidentally overturned the milk pails when he struck the Dragon with the yoke, and the milk poured out all along the Man's path.

I do not dare to comment upon this legend, as it seems to me quite perfect.

The Milky Way

[Otescu, and presumably his correspondents, routinely used the archaic term *Calea Laptelui*, "The Way of the Milk", rather than the more modern Romanian *Calea Lactee* for the Milky Way.]

In the Sky Myth, we saw an explanation for this

whitish band that divides the sky into two parts by its middle, and that it is seen well and completely only on the clearest of nights, without the Moon. Its appearance fills one's heart with a kind of superior respect, a kind of majestic feeling, similar to that experienced by a believer entering into a grandiose temple.

This majestic feeling was remarked upon by the Romanian people, who initially saw in the Milky Way the Emperor Trajan. Thus they called this whitish band *Troian* [6], or *Troianul Cerului* (the Trajan of the Sky) [terrestrially, troian was the name for snow-drifts] [13], or *Calea lui Traian* (Trajan's Way), or *Drumul Robilor* (the Slaves' Road), imagined as the way used by Trajan for his journey to Dacia, and for transporting the Dacian prisoners back to Rome, once they had been placed in the two Chariots, both Great and Little.

Those who only call this band the Slaves' Road have the following explanations about it:

* The Slaves' Road was marked on the sky by God to commemorate the numerous slaves captured by the Emperor Trajan in his wars with Dacia [8].

* In ancient times, the indigenous populace was transported as slaves into a remote Empire, which lay to the south and west (obviously, this is the Roman Empire). When they escaped,

they did not know the way to return to their country, so God indicated their direction on the sky [18].

* Trajan transported the captured Dacians to Rome, but they escaped, and ran back along Troian, travelling only at night [6].

However, these versions of the Slaves' Road were altered with time, and later became:

* A sign to be followed by escaped slaves generally [30].
* A sign for the conquerors, to transport away the enslaved people [4].
* A road for all the people when they reach the sky, because all are slaves on Earth [33, 34, 35].
* A road for those slaves who escaped from the eastern pagan territories [22]. For instance, on this road were said to have come back the Romanians from Moldavia, escaping from the Tartars. The Tartars' headquarters were in the Crimean Peninsula. Sometimes the Tartars were called *Căpcăuni* [a type of dog-headed monster-men]. A similar situation occurred for the Wallachian Romanians, escaping from the Turks, who came from the south. Thus, using this road, the escaped Romanians were able to return to their own country. It is said the Slaves' Road will disappear when slavery dies on Earth.

* The Slaves' Road is the way to be used by the people when they go to their Final Judgement, because all men are God's slaves [33, 34, 35].
* Using this road, the souls of the dead ascend to Heaven [12, 37].

In Greco-Latin mythology too, the Milky Way was the road to the palace of Zeus (Jove), used by heroes to enter the sky. The houses of the principal gods were situated to its right and left.

Egyptian poetry greets in the Milky Way the ethereal road leading to the house of the gods. As Cammile Flammarion says:

Even now, the Christians think they discovered in it the road of souls to the mysterious regions of eternity.

Concerning the name of the Milky Way, the Romanian peasants have other legends (similar to that where the milk pours out from the Shepherd's pails in the Sky Myth). I remember one of them told to me by my mother, in which the Milky Way was the milk poured from a mother's breast as she searched for her son in the sky, after he had been abducted by an Eagle. This legend is like that from Greek mythology, in which Hera, by Athene's counsel, gave her milk to the infant Heracles, after she had discovered him in a field where his mother Alcmene had left him, when she fled in fright at Hera's coming. The young Heracles sucked very intensely, causing so much milk to gush forth from Hera's breast that it formed into this Way in the sky, in which the stars are the drops of the goddess' milk.

Another Romanian legend [1]:

Once, a man stole some wheat straw from another man, and carried it off in a chariot. He was caught and brought to judgement, but he committed perjury, saying he was not guilty, although the victim had followed the traces of straw all the way to the thief's house. But God, to indicate which man lied, and to teach everybody forever that it is a great sin to swear falsely, put in the sky the appearance of the chariot, and the road which the straw had fallen out onto.

According to another version [21], the thief was the very godson of the person who suffered the loss, and the stolen straw was taken away in a basket. So this version does not include the chariot and the cart from the sky.

The last names for the Milky Way here are *Crângul Cerului* (the Sky's Grove) [37] and *Drumul Orbilor* (the Road of the Blind), but the second one is derived from a phonetical confusion with *Drumul Robilor* (the Slaves' Road).

CHAPTER FOUR:
THE SUN

For the Romanian people, the Sun is a living and holy being: Saint Sun [24]. A tale from Neamț County runs:

It is a sin for men to speak ill of the Sun, because if he is annoyed, he will no longer send light and warmth to them. Sometimes, he may even burn them!

The Sun is portrayed as a beautiful young man with a face so bright that the entire Earth is illuminated by it [10, 22]. Daily, the Sun traverses the canopy of heaven, riding a buffalo from early morning to noon (because this uphill way is the more difficult), a horse from midday into the afternoon, and a lion from the afternoon into the evening [9, 10]. Other peasants [10] simplify this to just a buffalo or a bull from morning to noon, and a lion or a horse from noon to evening. It is possible that this belief is connected to the fact that in the summertime, the Sun rises immediately after the constellation of the Bull, and when it sets, the constellation of the Lion (or the Horse in its popular Romanian form) is the first seen in the west.

At evening-tide when the Sun dismounts, he eats a piece of wafer, and drinks a glass of wine [10]; other peasants [9] think that he eats three times a day, eating well in the morning and in the evening, while at noon he has just the corner of a wafer with a glass of wine.

After his supper, the Sun travels through "other space", tărâmul celălalt, which surrounds the Earth, on foot, and reappears next day in the east. His unseen way between sunset and sunrise passes through underground caverns and hidden vaults [38; see the note for tărâmul celălalt on page 34].

Some peasants, however, think that the Sun rides only a lion to cover his whole route through the sky [3]. In the evening, tired, he and his lion sleep in a dark place. During their rest, some large monsters transport them back to the east [3]. Next morning, if the Sun slept well, he rises glad and bright, but if he slept badly, he rises angry, and hides his face among the clouds.

The lion does not like to stay near us all the time, so in the winter months, he runs further away from the Earth, to make us ask him to come back closer again [3].

Here is the peasants' physical explanation for the Sun [3]: He is formed from hot, bright beams, and because of this, he lights and warms the Earth; or he is a globe surrounded by beams that he sends to us through God's power.

These beliefs are not far removed from those of the ancients.

To the Greeks, Helius-Hyperion, the brilliant one who walks above, emerged in the east from the great depth of the River Oceanus, or from a lake formed by this river, and ascended slowly, as if on a hill into the sky until he reached the middle-sky, or culmination point. He then descended into the west, touching the Earth again as he sank into Oceanus, according to Homer.

Homer and Hesiod did not say how the Sun then covered the distance from the western Oceanus to that in the east, but the later poets told that the Sun after setting slept in a golden bed with wings, one of Hephæstus' (Vulcan's) creations, which transported the Sun on the River Oceanus (which surrounded the Earth) from west to east.

In the Rig Veda, the Sun's disc is compared to a fire-wheel that runs across the sky. This beautifully simple idea was developed by the Greeks in their poetry, where the wheel became a chariot drawn by luminous horses which breathed light and flames. Helius became the grandiose driver of this supernatural chariot. He was a strong, young, beautiful god, with a golden helmet and a glorious mantle, long hair and terrifying looks, who sent rays of light in all directions.

The Romanian peasants transformed the bed with wings into another bed carried by monsters, and the River Oceanus into a road surrounding the Earth. Helius remained a model of masculine beauty, but only his face radiated solar light. The peasants abandoned the superb celestial chariot, making the Sun instead a rider. In exchange, the fiery chariot and horses became those of Saint Ilie (Elijah), patron saint of thunder and lightning (as we will see in the chapter on atmospheric phenomena [Chapter Eight below]), dispossessing Zeus (Jove) of these elements. There is in fact a great similitude between the names Helius and Ilie.

Mythologically, in the winter, Helius goes to visit the happy Ethiopians, like Apollo to the Hyperboreans. Meanwhile in our popular belief, the unhappy Romanians are left to beseech the Sun to return, because he does not like to be with us all year long; but the Sun is indifferent, following his regular pattern in time, like a migratory bird.

CHAPTER FIVE:
THE MOON

Like the Sun, the Moon is a saint, a virtuous woman who lightens the night, and reposes during the day [22]. The Romanian people imagine the Moon as a beautiful maiden, the twin, or younger, sister of the Sun [3]. This familial link, and the fact that these two heavenly bodies do not meet for long in the sky, has given birth to many legends from the same source, namely that the Sun wanted to marry his sister the Moon, but that this wedding was stopped by something, and then the Moon fled from the Sun. The following two tales illustrate this.

The Sun [9], the most handsome of young men, could not find a more suitable wife for himself than the Moon, the most beautiful of young girls. But at their wedding party, the hedgehog heaped up dust-piles in the Sun's court, saying that these were fodder for the guests' horses.

The wedding guests were amazed, and asked the hedgehog, "Why do you do this?"

"Because if a brother marries his sister, God will stop the rain and all the food for the horses will disappear", answered the hedgehog, "so the horses must learn to eat dust".

Then the guests told the Sun what the hedgehog had said. The Sun became scared, and called off the wedding. And since then they never met again, because God separated them.

And about the hedgehog, because he did this good thing, it is said that it is a sin to kill him.

As we see, in the people's imagination, the hedgehog is a kind of philosopher, very modest and moral.

The Moon [3, 32] is the Sun's sister, but, because the Sun wanted to take her as his wife, she always ran away, so as not to be caught up by the Sun.

In Romanian popular beliefs, sometimes the Moon was said to be originally as luminous as the

Sun, because they were twins. The Moon's light became less, because after the failure of their wedding, as God wanted the Moon to keep out of the Sun's view, she hid at the bottom of the sea [10]. The sea water quenched her brightness.

For the ancient Greeks, Apollo, the solar god was also the twin brother of Artemis, the lunar goddess, the most beautiful and pure maiden, the goddess of clear light.

On the Isle of Crete, Artemis is mixed up with a local goddess called Britomartis, the sweet maiden, and as the Cretans' influence over the Mediterranean seacoasts was considerable, Britomartis became a second lunar goddess, of the seas, the Moon with the vaporous light, while Artemis remained the bright Moon of the land. Thus they became sisters.

Similarly, Minos, the principal Cretan god, was also mixed with Apollo, becoming his brother, and so also Zeus' son. This led to the generalization of the Cretan myth of Minos' love for Britomartis, as with the Romanians, but compounding Minos with Apollo and with the Sun, Britomartis with Artemis and the Moon. Here is that myth:

Minos, attracted by Britomartis' beauty, fell in love with her, and wanted to abduct her, but she ran off. When Minos was close to catching her, Britomartis threw herself into the sea from a high rock, landing in a fisherman's net. Then Artemis made her into a deity as a reward for her chastity. From that time forth, Britomartis oversaw the seas at night, appearing to the people of the shores and isles.

Descharmes says, "Graceful personification of the Moon, directing the sailors in their travels, illuminating the activities of the fishermen by night, and, reflecting herself in the surface of the waves, playing among the nets. Or, in accordance with her legend, she is the image of the travelling Moon who walks over the forests and mountains, but who runs from the Sun who desires to catch up with her, and ends by plunging into the abysses of the seas. In a word, she is the lunar Artemis of the Sea-peoples."

Her cult circulated the shores of the Mediterranean, from Crete to Marseilles.

So this cult was also present in Italy, and the Roman legions brought it to Romania, where the legend survives to this day.

The Moon's phases impressed the Romanian peasants too, but they did not understand what they were, and so they found a simple Christian explanation [38]: Every month, God renews the Moon to demonstrate that He can make little things great, and great things small.

When the new Moon first appears, the peasants look to see if they are carrying any money in their bag. If they are, it will be a good month. Some of them take a coin from a bag and make the sign of the cross while looking at the new Moon, asking her to increase their money.

The peasants also see the Moon's spots and think, without understanding that they are mountains and valleys, that the Moon is a man's round face, with his eyes and nose formed by the spots.

Most of the peasants, however, explain the Moon's spots as being in essence the shepherd called Abel (the first shepherd on Earth, according to the Bible), who was killed by Cain. Here are some other explanations:

* The Moon's spots are a shepherd with his flute and sheep. His name is Abel, and God set him on the Moon to show mankind how mild he was [3, 32].
* The Moon's spots are two brothers. The big brother killed the little one, and as a punishment, has to carry his dead brother's body, which drips blood [18].
* The Moon's spots are two brothers, one of whom is punished by having to drink coal-oil from a wooden bowl because he killed the other [37].
* The Moon's spots represent Cain, who drinks the blood of his brother Abel, whom he killed [24].
* On the Moon can be seen a shepherd with his

sheep, propped up by his crook; near him is a fountain and a woman, who came here to drink the water and speak with the shepherd [10].

This last image concerning the Moon's spots is the most widely-circulated. It may be a projection onto the Moon of the constellation of the Shepherd, who has beside him in the sky the Great Maiden with a Yoke and the Fountain of the Crossroads.

CHAPTER SIX:
ECLIPSES

Darkenings of the Sun and Moon are called eclipses.

The Romanian people think that eclipses of the Moon exist because the Moon is eaten by unearthly monsters called *vârcolaci* ["worm-like creatures", from the Latin *vermicolacius*, varkolak in Slavonic].

The vârcolaci are explained in various ways as being:

* a kind of animal smaller than a dog, or simply small dogs;
* balauri or zmei [24, 37];
* animals with many mouths like octopus

suckers [32];

* ghosts called *pricolici* [8];

They also have various origins, including appearing:

* from children who die unbaptized [30];
* from children born to unwed parents [21];
* if someone touches the fire with the wooden stirrer when making porridge [37];
* if someone brushes dust towards the Sun while sweeping their house at sunset [37];
* if a woman spins without a candle in the night, especially at midnight, and especially for the purpose of making spells with the thread thus spun. Werewolves come to those threads; and the threads work themselves into a werewolf's life-path. As long as the threads do not break, the werewolves staying on them are strong, and can go where they want: then they can attack the heavenly bodies, tearing them with their teeth; and so they also tear at the Moon, which sometimes they turn all to blood. But if the thread to which the werewolves are connected breaks, then they lose most of their power and must flee away through the air [38].

The important question is, how does the Moon remain whole if she is eaten by the vârcolaci? Some people [17, 38] explain this by saying that as the Moon is more powerful than the vârcolaci, she can only be bitten by them, because if she were to be swallowed, it would be the end of the world. Thus many peasants make a terrific din when there are eclipses, in order to scare away the vârcolaci; they ring the church bells, hit trays, shoot guns, play instruments loudly, and so forth.

Other peasants say that because the Moon is so great, she tires out the vârcolaci which attack her, forcing them to liberate her. Still others say that it can all be explained through running; the Moon runs very quickly, and the vârcolaci can only nip briefly at her, before falling behind.

Eclipses of the Sun are explained by some peasants just like lunar eclipses; the vârcolaci eat the Sun, and the Sun is saved because the lion which he rides fights with the vârcolaci, protecting him [3].

An alternative explanation is that during a solar eclipse, the Moon runs quickly past the Sun, and God darkens the Sun in order that he should not see the Moon, because of their unacceptable love [9]. This belief is actually quite close to reality, because during a solar eclipse, the Moon passes in front of the Sun, which is the cause of the eclipse.

Some peasants say that the Sun, seeing men's wickedness, is disgusted, and hides his face [18]; or the Sun's darkening is a sign from God to cause men to repent their sins [24]. This idea also recurs in an extended form, since there are peasants who think that eclipses of the Sun and Moon happen when God commands the vârcolaci to eat them, in order to scare people into abandoning their sinful ways [22].

Here is a further explanation, with a little poetry to it [10]: As Saint Sun and Saint Moon are very beautiful, some monsters that live in the sky desire to kiss them. When they try to do so, however, they actually swallow them, but the Sun and Moon are able to draw out of the monsters' mouths, and return more beautiful than before. A more prosaic explanation also advanced [33, 34, 35] is that the vârcolaci gnaw at the Sun and Moon so they can harm us on Earth, but we are very fortunate because the Sun and Moon can run so quickly, and avoid damage.

When eclipses, especially of the Sun, occur, some Romanian peasants consider it necessary to light Easter candles and pray [30].

As we see, although the peasants are Christian, in their beliefs still exist reminiscences of the ancient Romans, who perceived the Sun and Moon as two deities. Otherwise, why would the Holy Sun and the Holy Moon be Saints? The belief that the Sun's disappearance would mean the world's end has its own logic. The most impressive form of this terrible event is the poem "The Dream" by Lord Byron. However, it is easy to imagine what the Earth would be like without the Sun's light and heat. The Romanian people think the same catastrophe would also follow if the Moon failed to light the night, because she is the Sun's sister, and we know that the Moon is seen only because she is illuminated by the Sun, so the end of the Moon's light would indeed precede the end of solar light, as Lord Byron observes in his poem.

I shall conclude these Romanian popular beliefs about eclipses with one more [35]: In a war, if the Sun darkens, the omens are bad for the Christian race; but if the Moon darkens, it is a bad sign for the Ottomans. This belief recalls that of Alexander the Great at the Moon's eclipse on the eve of the battle of Arbela.

Connected to these ideas are those of the peasants who think solar and lunar eclipses foretell wars and calamities [11], but they see these as more especially foretold by comets.

One of the writers about the Indo-European race's origins, Adolphe Pictet, was surprised because he could not find any written mentions of eclipse myths among those of the ancient Greeks and Romans. It seems certain, however, that their beliefs were similar to the Romanians, because they made the same great din during eclipses. Thus Tacitus in his Annales, and also Juvenal, comment upon such noises being made during a lunar eclipse. A further old belief concerned the Thessalian witches, who praised themselves as they claimed to be able to save the Moon from the dragon who wanted to swallow her in an eclipse. The Thessalians used their magic, or made a terrific noise by hitting pails, to accomplish this.

The belief that eclipses represent an attack by monsters on the Sun or the Moon is universal, however, and was so from the earliest times, even with the ancient Aryans. In the Mahabharata, the old Aryan poem, the demon Rahu surreptitiously drank of the gods' elixir of immortality, but was seen doing so by the Sun and Moon, who told this to Vishnu. Vishnu beheaded Rahu in his anger, but Rahu's head was immortal, and it still follows the two great lights in the sky, trying to swallow them. Naturally, when this fiendish head swallows the Moon or the Sun, the body in question is able to escape again through Rahu's open neck; this is an amusing explanation for the integrity of the Sun or Moon after an eclipse. The same story is found in the Vishnu Purana. Later, this myth travelled from India to Mongolia, where Rahu became Araho, and the Mongolians too made a great noise during an eclipse to drive the monster away. Rahu could easily have become "draho", and been mixed up to form the Greek "drakon" and the Latin "draco", becoming "dragon", the Romanian balaur.

So the Romanian belief that the vârcolaci were the balauri clearly has a very early origin. However, the Romanian etymology of the word vârcolac indicates it derives from "wolf", thus the little dogs should actually be little wolves.

This belief about the wolf-vârcolaci must be very old too, and probably also Aryan, since it is found even among the Scandinavians.

CHAPTER SEVEN:
COMETS

The Romanian people call comets *Stele cu Coadă* (Tailed-Stars), and think they are divine signs placed in the sky to announce the coming of great hardships, such as bloody wars between kings and emperors [24], epidemics or plagues of men and cattle, famines, etc., and that these will occur in the direction the comet is visible [26, 30]. Even the end of the world will be heralded by a tailed-star, thus many peasants think, on seeing a comet, that the end of the world is at hand, and prepare themselves for death using their remaining time for fasting and prayers [8].

In fact, all the beliefs in the evil portents of comets existed from the earliest times with all peoples. What can we say when, even in our own times, many prognosticators forecast that a comet will destroy the Earth, and give the time of its coming? But the cunning comet does not wish to respect their anticipations!

Romania's war for independence in 1877-78 contributed to the people's beliefs in the forecasting ability of comets. In a paper which I received from Vâlcea County, a peasant remarked [3]:

Sometimes, stars are seen with tails, which forecast bad events, like wars, plagues, etc. In 1877, while the Romanians fought with the Turks, a star was seen in the sky with a little tail; then the tail began to lengthen, and the star drew nearer to us; only close to the end of the war did this tailed-star disappear.

CHAPTER EIGHT:
ATMOSPHERIC PHENOMENA

The Romanian peasant has a few ideas about atmospheric phenomena.

The Wind

This is formed from the breath of evil monsters who want to destroy all the things on the Earth, so that only they will remain. Of course, this belief refers just to strong winds, the ones which dry things up in summer (*Austrul*), and freeze them in winter (*Crivățul*), because the lesser wind is considered useful in some cases. The following story illustrates this [38]:

One day, a beautiful and intelligent girl met three young men, who said to her, "Good morning, beautiful girl."

The girl, who recognised the three boys as Wind, Heat and Frost, answered, "Thank you to one of you three!"

The three boys were confused by this. Who was the one?

"Me", said Heat, "because she knows that, if I want, I can stifle her during the agricultural works of summer."

"Me", said Frost, "because she knows that, if I want, I can freeze her on the road in winter."

"I think it will be better if we ask her", said Wind.

All three agreed to this idea, and they ran after the girl to ask her which was the one she had meant. "Wind", answered the girl.

"I'll catch you in the summer", menaced Heat.

"I don't care", replied the girl, "if only there is Wind."

"I'll catch you in the winter", menaced Frost.

"I don't care", replied the girl, "if only there is no Wind."

In the magazine Șezătoarea, second year, under the title "The Wind", are the following lines:

The Wind is a young, handsome man, whose house is in the air. He lives with his mother and is unmarried. He would like to marry, but he cannot, because he has not found a wife good enough for him. When he grows tired of racing round the world looking for a suitable wife, he stays in his house, and throughout the world there is silence. When his desire for a wife grows strong again, then the Wind reappears in the world once more.

The souls of men climb to Heaven and Hell on the wings of the wind, until they reach the limits of the air.

The zmei also travel on the wings of the wind.

When the wind blows powerfully for a long time, it is because he bewails the soul of one who has been hanged or drowned.

The wind does not blow in Heaven.

For the ancient Greeks and Romans, there was a god of the wind, Hermes (Mercury), who was also the god Psychopompus or Psychogogus, the god who conducted the souls to their places after death.

Another belief the Romanian people hold concerns the dust-filled tornado, or dust devil, formed by the wind, the whirlwind. The dust devil is very dangerous, because it hides a devil in its midst. This devil can pull people up into the sky and then let them fall on the ground, to be crippled. He can also cause paralysis in anyone caught in these swirling winds. "He takes a hand or a foot, or twists the mouth", say the peasants.

The same effect can also be made by the Iele, but the tornado is replaced by their Hora, the ring dance.

Iele, Dânse, Șoimane, Dobre, Sfinte, or especially Joimărițe are names for a group of malignant fairies, very beautiful, and attractively dressed, but who will not forgive anyone who sees them [5, 6]. They live in the air, but by night they come down to the Earth, singing and dancing. Where they touch the Earth, the plants grow pale and die beneath their feet, becoming like ash from under a fire, and that place becomes devastated. "God forbid that you should meet them!" There is a general belief that if they call your name from behind, and you turn to see them, they will disfigure you. They also disfigure those who sleep outdoors in the night, dancing a Hora around such imprudent persons.

An interesting case is that where the Joimărițe punished a slothful girl who did not spin hemp until Joia Mare [Joi = Thursday, Mare = Great, so "Great Thursday", Maundy Thursday, the Thursday before Easter]. The term Joimărițe proceeds from Jove (Jupiter) + Marte (Mars), but in time and with usage by the peasants, the name of the god Marte here became shortened, and its sense altered, to Mare. The real explanation of this belief stems from the fact that a Romanian peasant-woman contributes to the agricultural work alongside her husband; so her winter work, which includes spinning, must be completed before the agricultural field work begins again, which is usually immediately after Easter. Only during the autumn do the peasant women have time for flax and hemp spinning.

These malignant fairies must be reminiscences of the wind deities, the Harpies, the death goddesses, the Dryads, and the nymphs who protected the forests, and who punished those who cut into the great trees under their protection; such goddesses performed ring dances and sang songs around their protected trees too.

There are Romanian peasants who have been out walking in the night through forests or fields, and thought they heard the aerial song of the Iele, just as the ancients thought they heard the Dryads' song. Flutes placed in tree-hollows near where the Iele sing and play become magic flutes. If anyone plays them afterwards, those nearby are forced to dance, and if the flutes are played for a long time, the dancers must dance on until they die.

It is possible that the name Iele derives from Elo (Aello; Aellopus), the Harpy representing the breath of the hurricane.

The Rain

This is given to the Earth when God wishes to do so. When He does not, God gives drought. God also makes the clouds. Hailstones are made by the Solomonari, who pass through the clouds riding the balauri, carrying the hailstones [30]. The Solomonari throw hailstones wherever they want to. It is possible that the small, mobile clouds (which the Italians call Ascitizi) [Stratus fractus, Cumulus congestus arcus or Cumulonimbus arcus] that appear before the stormier clouds arrive, especially before the hail-stone storm clouds, could have given the peasants the idea of the Solomonari who ride the balauri and choose the place where the hail will fall.

Concerning the name Solomonari, I think it comes from Solomon's wisdom. The peasants do not know exactly who Solomon was, but they think he was someone quite extraordinary. They also believe that Solomon was someone who knew how to dominate evil spirits. Some of these spirits were once trapped by Solomon in bottles, whose corks were sealed by him, and then the bottles were thrown into the sea. The evil spirits saw Solomon's seal on their bottles, and did not dare to emerge again. Other spirits were cast into the Earth's depths in a similar way by Solomon. Only a few spirits were not thus enclosed, and they remain free, causing a lot of trouble both on Earth and in the sky. The Mahomedans held these same beliefs about Solomon, but they expressed them in more elaborate, wonderful stories, although without any connections to natural phenomena, unlike the Romanian myths.

The Lightning, Thunderbolt and Thunder

These are the work of Saint Ilie (Elijah).

In the peasants' beliefs [9, 38], Saint Ilie follows the evil spirits and the Devil in the sky, moving across the clouds accompanied by his horses, chariot and a fire-whip. Thunder is the clatter of the horses' hooves and the rolling of the treasure-cart's wheels. It occurs when Saint Ilie's horses run too quickly through the clouds. When Saint Ilie whips his horses with the fire-whip, then lightning appears. This makes the horses dash along, which means the thunder follows immediately after the lightning. The thunderbolt [lightning strike] happens when Saint Ilie shoots at the devil with an arrow. Being thus pursued, the Devil hides where he can, in men, cattle, houses, trees, etc., but Saint Ilie sees him wherever he is, and thunders after him. However, the thing in which the Devil hides is in danger too, so when lightning is seen, the people make a Christian hand-cross over themselves, in order to remain free from the Devil [30].

Some peasants think that when it thunders, the bulky dark clouds are fighting, butting at each other with their heads. From such a fight, a type of foam falls on the Earth which is good for wounds [33, 34, 35].

CONCLUSION

Christianity did not erase from Romanian thought all the old beliefs and legends from the pagan Romans. From them, some tales remained clear, but others became confused. This happened with all the Christianized peoples, however.

These old beliefs are included in the people's innermost convictions, but they are not annexes of the Christian religion.

Some indeed are heresies, which hinder the peasants in their field work on days which are not Christian holidays; others make the peasants superstitious; still more facilitate all kinds of frauds with which the poor peasants are robbed from time to time by lazy, wicked and sly people, such as the

Solomonari and sorcerers in general; and there are also other evils, but they have a positive aspect, telling of the people's origins.

Such heresies, and the people's legends and language, together with their history, form the Golden Book of that people. Then, even history must recognise them, but this source is frailer, needing more discrimination to use. From this work, we have seen so many valuable data about this Golden Book of our origins, that nobody could be in any doubt about our Dacian-Roman beginning. Thus, with our heads held high, we can say like Ovid: *Et documenta damus qua simus origine nati* ["And we give proof of the origins we're born from"].

LIST OF SOURCES USED

[In the following list, place-names, including the appropriate Romanian county, precede the names of the teachers involved in collecting the information, followed by the names of those contributors who provided what Otescu considered to be especially valuable information. Note that the final three entries in this list, numbers 40, 41 and 42, did not appear in Otescu's own version. We discovered the places among his footnotes, and have added the locations here, although unfortunately, we know nothing of the contributors from those areas. Curiously, we also discovered that entry 7 in this list, Costeşti in Muscel, was not mentioned in Otescu's text notes, only in this final list. - Translators.]

1. Almajelul, Mehedinţi. The priest I. Gomoiu.

2. Buteşti, Vâlcea. I. D. Brănişteanu.

3. Stăneşti, Vâlcea. Const. Rădulescu.

4. Băsenii-Stârci, Argeş.

5. Măcăi, Argeş. P. Dumitrescu.

6. Scheiu, Argeş. Ion Sinescu.

7. Costeşti, Muscel. C. Becuţ.

8. Braniştea, Dâmboviţa. C. Ionescu.

9. Ezerile, Prahova. N. Bordeanu.

10. Călugăreni, Prahova. Iorgu Ştefănescu.

11. Talea, Prahova. R. N. Hoita and the shepherds Gheorghe Nedelcu, Ion Baciu and Gheorghe Şerbănoiu.

12. Mărginenii-de-Sus, Prahova. Naie Popescu.

13. Scorţeni, Prahova. G. Stoicescu.

14. Ruşavăţ, Buzău. Ana Morareţ.

15. Dumitreşti, Râmnicu Sărat. I. Dumitrescu.

16. Drăgăneşti, Olt. C. Pănculescu.

17. Gura Boului, Olt. Ion Drăgulescu.

18. Ciocăneştii-Mărgineni, Ialomiţa. Em. Popescu.

19. Gaiţa, Ialomiţa. Anghel Vlădescu.

20. Ulmu, Brăila. G. Popescu and the old man Mircea Pârlog (96 years old).

21. Turcoaia, Tulcea. C. Popescu Măgureanu and the old men Vlad Lupoaica and Ion Tufeci.

22. Răstoaca, Putna. G. Cursaru and the peasants Gheorghe Angheluţă, Vasile Zamfir and Vasile Tarbă.

23. Grigoreni, Bacău. I. Ghindeanu and the old men Vasile Jigău (105 years old), Gavrilă Sava, Gheorghe Papuc and Ion Şerban.

24. Negreşti, Neamţ. D. A. Alexandrescu.

25. Cătunul Mănăstirea Bistriţa, Neamţ. Gh. Constantinescu and the old men Andrei Ignat (95 years old), Ion Roşu, Toma Ardeleanu and Costache V. Gavril.

26. Călugăreni, Neamţ. D. A. Locşa.

27. Dobreni, Neamţ. Darius and the old man Constantin Boboc.

28. Măstăcani, Neamţ. A. Baciu and the old men Vasile Gr. Ungureanu and Ion Barcan.

29. Uscaţi, Neamţ. I. Dumbravă and the old men Vasile Davija, Constantin Carhiţă, Vasile Florian, Vasile Mazarinu (90 years old) and Ion Lupei.

30. Bârcu, Neamţ. G. Popovici and the old men Pavel Ciobanu, Gheorghe Boiţă and Dumitru C. Ciobanu.

31. Cracaoani, Neamţ. G. V. Nicolau.

32. Cracăul-Negru, Neamţ. C. Puntescu and the old men Gheorghe A. Antăloaii and Ion Subţirelu.

33. Gura-Hangului. G. Leonescu.

34. Grinţieş. C. Cavrileşteanu-Taşcă.

35. Bistricioara: P. Cădere. In entries 33-35 all of the places are in Neamţ county. Also the old men Toader Lăcătuşu and Ştefan A. Savi should be acknowledged here.

36. Hangu, Neamţ. D. Diaconiţă.

37. Cristeşti, Botoşani. I. Iordăchescu.

38. Lişna, Dorohoi. I. Cernat and the old men Gheorghe Drâmbă, Dumitru Printrinjel, Manolache Frunză (95), Dumitru Lupan and the old women Maria Anton Antonecioaia and Zamfira Bilibori.

39. Cordăreni, Dorohoi. I. Balinescu and the peasant Tudor Simion.

[40. Odăi, Buzău. Unidentified correspondent not listed by Otescu.]

[41. Suhaia, Teleorman. Unidentified correspondent not listed by Otescu.]

[42. Verneşti, Buzău. Unidentified correspondent not listed by Otescu.]

TRANSLATORS' AFTERWORD: ON THE FATE OF ION OTESCU'S BOOK

Forgotten for several decades, Ion Otescu's book was rediscovered after 1990 by Dănuţ Ionescu, who together with Erika Lucia Suhay and Monica Ciobanu reported on its content in the astronomy magazine "Orion" in Bucharest.

After 1995, Alastair McBeath and Andrei Dorian Gheorghe started translating it into English, using it also as a source of inspiration for a series of articles about Romanian starlore that they published in "The Dragon Chronicle" (then the international journal of dragonlore), where Andrei also published some astromythological poems on the same theme. For example, in regard to Romanian starlore, Ion Otescu and his team of teachers and priests from the countryside, identified 39 definite and 3 uncertain constellations in the Romanian sky, but the long, straggling line of often faint stars comprising Hydra was not among them. Although one of the largest modern constellations, probably this very size, length and faintness meant it did not interest the Romanian peasants for their guidance in working the fields. In trying to continue Otescu's work, the translators found something that could be connected to Hydra in the old folklore tale "Tinereţe fără bătrâneţe şi viaţă fără de moarte" ("Youth without Old Age and Life without Death"), published by Petre Ispirescu in 1862, in which the hero, defeated two monstrous flying creatures, *Gheonoaia* and *Scorpia*, who could not live together. The first caused violent spring storms, while the second caused summer fires. This could suggest Gheonoaia as the constellation Hydra, with Scorpia as the constellation Scorpius, a hypothesis demonstrated in 1998 in an article published in "The Dragon Chronicle".

It is worth noting too that the Romanian star-patterns Otescu was able to identify, included differences in what they represented even in places quite near one another sometimes. There are also cases where several constellations were seen in what are modernly (since the official list of constellations was published by the International Astronomical Union in 1928) just one, such as Taurus, where the Pleiades and the Hyades can appear as independent constellations, or Orion, divided into up to four constellations, or where modernly multiple constellations, including Serpens and Ophiuchus, or Andromeda and Pegasus, were formed into one.

After the publication of the English translation of the book online in 2009, the Baia Mare Planetarium started the Traditional Romanian Constellations Project (institution coordinator Ovidiu Ignat and project manager Mircea Liţe), with this book as their main source of guidance. Thus, this team published an informative brochure, created a digital exhibition and realized the Romanian version of the sky for the Stellarium application, for which they used this translation. The following Addenda chapter has further details on this Project.

Finally, by publishing Ion Otescu's book in a language of wide international use, we hope it will become better appreciated as an important part of the world's astromythological treasury, given that it represents a spiritual Romanian Mount Olympus, whose summit is the Sky Myth, in which good defeats evil on a stellar level. Hence we end with an astromythological poem inspired by this tale, which was published in "The Dragon Chronicle" in 1997:

ROMANIAN COSMIC LEGEND

By Andrei Dorian Gheorghe

Two armies in the sky
and the Man defeating
the Devil.

Two armies in the sky,
and the Shepherd defeating
the Dragon.

Two armies in the sky,
and the Milky Way defeating
the Darkness.

ADDENDA:
THE TRADITIONAL ROMANIAN CONSTELLATIONS PROJECT

Stars, meteors, planets, galaxies, comets, the Moon... What could be more uplifting and relaxing than looking at the sky on a clear night? Humans have looked up to the sky since the beginning of time. Some to communicate with the deity or deities they imagined to dwell there, others to discover and learn the laws of nature. Or simply because the heavens fascinate and awe us with their majesty. In essence, these are just different ways in which we seek answers to more delicate questions: Who are we? Why are we? Where do we come from? Why on this "ship" traveling around this Sun? Where are we heading?

Primitive man, in order to organise his daily activities more efficiently, had to read time in the sky by patiently watching the stars. By following the movement of the Sun in the sky, he learned to predict the growing seasons, building calendars which he could use to plan his agricultural work. Thus astronomy, if not the oldest, certainly one of the earliest, natural sciences, emerged as a necessity for everyday life.

Every nation formed its own image of the celestial vault, according to its needs and development. In the beginning, the Romanians were a people of farmers and shepherds, who used the starry sky for orientation in time and space. Our ancestors watched the stars and formed their own representation of the sky.

Ion Otescu's work, published in 1907, is an extraordinary exercise in documenting the Romanian constellations as seen by our ancestors. The research was done exclusively in writing, the mathematician Otescu sending letters to various places in the country, to priests and teachers, asking them to send him back what they had discovered about what the locals knew of the sky, stars and constellations. The answers obtained were systematised and presented in his publication. Just taking into account the immense effort made by the distinguished school teacher

Otescu in collecting and processing the data (at the beginning of the 20th century!), we can appreciate that the work is of exceptional value. We are also talking about 39 Romanian constellations identified, many of them coinciding with the official ones, but we have 19 "innovative" constellations, which have no connection with the official ones.

As Otescu's work rippled through the ages, it regained popularity more than 100 years later in a small city by the mountains of northern Romania. It was the main source of information for the "Constelații Românești Tradiționale" Project ("Traditional Romanian Constellations" Project), started by the Baia Mare Planetarium in 2012, and whose story continues today through an exhibition created thanks to this effort. The aims of the Project were to bring to light the complex origins of the star myths, the structure of the sky and the connections between earth and sky for the traditional Romanian villagers, in a modern, digital museum perspective, in order to attract younger people, who have the right to know about the preserved treasury of popular knowledge regarding sciences such as astronomy and meteorology to be found in connection with the old Romanian constellations and their association with Romanian stories and legends.

In the words of our dear departed friend, Mircea Lițe, the Project Manager: "the Romanian celestial vault says more about the Romanian people than do tomes of books. We find here Greek and Roman influences, but also unique aspects specific to the Romanian people, a Christian people of Dacian and Roman origin". If the "official" sky predominantly constructed by the Greeks (the official constellations of the International Astronomical Union) abounds with ancient deities and heroes, those of the Romanians are dominated by pastoral agrarian, religious and historical constellations. We can therefore classify the Romanian constellations as divided into three broad categories: a pastoral agrarian, ancestral component (such as the Scythe, the Broad Axe or the Auger), a religious component (God's Chair, the Great Cross) and a historical component (Romulus and Remus, the Archer). These three components intertwine harmoniously and form the specific Romanian aspect of the starry sky, and there is also a relative unity of mythological concepts, the Romanian people's vision of the sky being approximately uniform throughout Romania. Although, apparently, since Ion Otescu and Tudor Pamfile [1883-1921, a mythologist who tried to continue Otescu's work by writing "The Mythology of the Romanian People" and "The Sky and Its Ornaments"] no one has dealt with the traditional

Romanian constellations, interest in them has existed all this time.

Returning to the "Traditional Romanian Constellations" implemented by us at the Baia Mare Planetarium (www.planetariubm.ro, www.crt-ro.com, the name of the institution being currently Muzeul de Științe Astronomice / Astronomical Science Museum), it was carried out in three stages, all co-financed by Administrația Fondului Cultural Național (the Administrator of the National Cultural Fund). In the initial stage, the most important Romanian constellations were described, the legend of the Romanian sky was presented in a series of videos and the basic exhibition was organised at the institution's headquarters on TV screens. The second phase, which we called the "National Phase", included a collaboration with all the planetariums existing at that time in Romania, in which the description of all 39 Romanian constellations was finalized and turned into a mobile exhibition that has so far been presented at 14 locations in Romania (Miercurea Ciuc, Toplița, Satu Mare, Oradea, Negrești-Oaș, Timișoara, Hunedoara, Carei, Bucharest, Buzău, Cluj-Napoca, Sighetu-Marmației, Pitești and Golești). The number of visitors who benefited from the results of this project is estimated to be 500,000 (comparable to the whole population of Cluj-Napoca).

The next stage of the Project was setting up an exhibition for the blind or visually impaired. This is how "Cerul în mâinile tale" ("The Sky in your hands") Project came about, with a basic exhibition at our headquarters, adapted to the needs of the blind, consisting of tactile boards and audio-video information, and seven special planetariums for the blind (a set of two small domes for individual use, corresponding to the winter sky/summer sky, on which beaded constellations were placed). These seven special planetariums were distributed to the project partners.

Another important result of the Project was the introduction of Romanian stellar mythology following Otescu's material into a widely used virtual planetarium program: the Stellarium software. Thus, in addition to the many non-European mythologies present in this celestial simulator, since 2013 users also can find the constellations of the Romanian sky, as viewed by our parents and grandparents.

What if astronomical associations, planetariums, ethnologists, were to repeat the research carried out by Professor Otescu? But with today's means of communication! We could ask not only the elderly, but also extend the study to young people. Would we enrich the heritage of the constellations of the Romanian sky? Would it be poorer? Obviously we don't want answers "off the Internet". It is a topic that can be reflected upon, and we must debate it because we are dealing with a "treasure". Traditional Romanian constellations are an intangible asset of the Romanian people that must not be forgotten.

In the hustle and bustle of life these days, we seem to have lost the ability to look up to the heavens. Have we lost it just for the moment, or are we a lost people among other peoples who have bowed our heads to monitors and mobile phones? The problem is not just with us – it's global! Massive light pollution and fatigue accumulated over a hectic day are just two of the reasons we are out of tune with the splendour of the starry sky. And that's a shame. Because looking up at the sky, man has learned to master the Earth. Because we may not have much time left. It is quite possible that soon, the artificial satellites launched by mankind in order to have global internet without any censorship (X Space, Amazon, One Web and others) will create a "Brownian motion" of these "bugs" in the sky that will "scribble" over the sky incessantly, making it impossible to identify the constellations. Certainly, these thousands of tiny dots in the sky will be a symbol of technological progress but, unfortunately, they come with the destruction of virgin nature, the starry sky. Is it worth it?

Ovidiu Ignat, Director
Baia Mare Astronomical Science Museum
March 2024

ROMANIAN SKY MAP

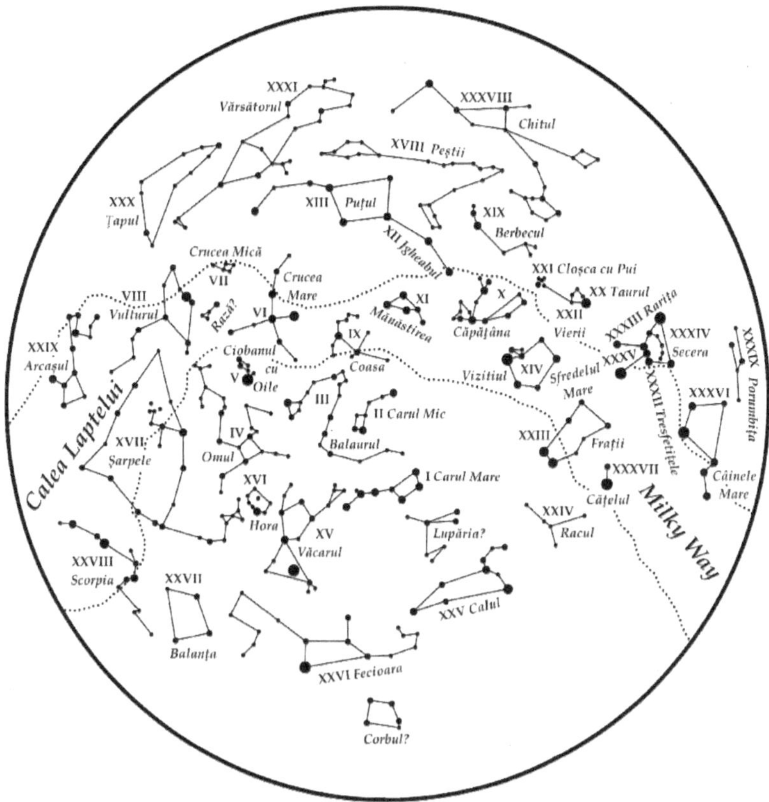

The Romanian starry sky, based on Otescu's 1907 star-map. Only some stars are shown, linked by imaginary lines. Each constellation has just a single name, and the Roman numeral used in the text. The general band of the Milky Way is shown too. The map illustrates the whole sky visible from Bucharest (and any site around 44½° North). Non-Romanian star patterns have been omitted, leaving blank areas, focusing attention on areas Otescu regarded as significant.